T0350270

Applications of Heuristic Algorithms to Optimal Road Congestion Pricing

Road congestion imposes major financial, social, and environmental costs. One solution is the operation of high-occupancy toll (HOT) lanes. This book outlines a method for dynamic pricing for HOT lanes based on nonlinear programming (NLP) techniques, finite difference stochastic approximation, genetic algorithms, and simulated annealing stochastic algorithms, working within a cell transmission framework. The result is a solution for optimal flow and optimal toll to minimize total travel time and reduce congestion.

ANOVA results are presented, which show differences in the performance of the NLP algorithms in solving this problem and reducing travel time, and econometric forecasting methods utilizing vector autoregressive techniques are shown to successfully forecast demand.

- The book compares different optimization approaches
- It presents case studies from around the world, such as the I-95 Express HOT Lane in Miami, USA

Applications of Heuristic Algorithms to Optimal Road Congestion Pricing is ideal for transportation practitioners and researchers.

Don Graham is a Data Scientist and Consultant. He served as data science professor at Northwestern University and transportation professor at Florida Institute of Technology for over ten years, in addition to working in industry at the Institute for Defense Analyses, Lockheed, and AVIS.

Applications of Heuristic Algorithms to Optimal Road Congestion Pricing

Don Graham

CRC Press
Taylor & Francis Group
Boca Raton London New York

CRC Press is an imprint of the
Taylor & Francis Group, an **informa** business

First edition published 2024
by CRC Press
2385 NW Executive Center Drive, Suite 320, Boca Raton FL 33431

and by CRC Press
4 Park Square, Milton Park, Abingdon, Oxon, OX14 4RN

CRC Press is an imprint of Taylor & Francis Group, LLC

© 2024 Don Graham

Library of Congress Cataloging-in-Publication Data
Names: Graham, Don (Data scientist), author.
Title: Applications of heuristic algorithms to optimal road congestion pricing / Don Graham.
Description: First edition. | Boca Raton, FL : CRC Press, 2024. | Includes bibliographical references and index.
Identifiers: LCCN 2023028326 | ISBN 9781032415659 (hardback) | ISBN 9781032415666 (paperback) | ISBN 9781003358701 (ebook)
Subjects: LCSH: Congestion pricing—Data processing. | Congestion pricing—Mathematical models.
Classification: LCC HE336.C66 G73 2024 | DDC 338.1/1—dc23/eng/20230705
LC record available at https://lccn.loc.gov/2023028326

ISBN: 978-1-032-41565-9 (hbk)
ISBN: 978-1-032-41566-6 (pbk)
ISBN: 978-1-003-35870-1 (ebk)

DOI: 10.1201/9781003358701

This book is dedicated to my wife Khandie and newborn daughter Amelia.

Contents

Introduction

1

1.1 REVIEW OF CONGESTION PRICING: DOMESTIC AND INTERNATIONAL IMPLEMENTATIONS

In this section of the proposal, we review the current state of the congestion pricing literature, which is a strategy for reducing the ever-present problem of traffic congestion and commuter delay in today's urban cities. Traffic congestion in the United States is a major source of financial loss for business and commerce due to lost time and productivity.

In the United States the ten most congested locations occur in urban regions such as Los Angeles, California; Houston, Texas; Chicago, Illinois; Washington, D.C.; Atlanta, Georgia; and Phoenix, Arizona. These ten locations by themselves result in a cumulative annual delay of more than 200,000,000 hours. Since we know that lost time equals lost money, we can see that there is a significant financial cost associated with congestion, which must be addressed.

When faced with scarce resources, governments often are unable to build new roads or freeways to meet all the transportation demand for their cities or urban regions and must find other means to improve efficiency of the road facilities, which serve their community. The method of congestion pricing is one such effective method, which has been applied in major metropolitan areas to successfully reduce congestion and improve throughput on highways. Some examples of locations where congestion pricing has been successfully implemented include London, Stockholm, Singapore, and US-91 in San Diego California.

In London, England, the problem of congestion was addressed by implementing a surcharge for driving a vehicle within the city perimeter during certain hours of the day. The driver who planned to use these city streets would purchase a pass for a predetermined price and thus gain access to these city locations during peak hours. The enforcement method used was a network of

DOI: 10.1201/9781003358701-1

1

cameras that would take photographs of vehicles registration when they passed certain points in the network. The date and time would be matched with a database of eligible drivers, and toll would be deducted from the driver's pass.

If an ineligible driver were using the city streets during peak hours, then their information would be recorded, and they would be issued a toll or citation. The implementation of congestion pricing in London was an immediate success in reducing traffic by 15–20% in the city. The drivers, which had been displaced due to the implementation of congestion pricing, were shifted in their mode of transportation to mass transit (subway tube) since ridership on buses and subway increased an equal amount as the reduction (15%) on highways.

In Singapore, the downtown area adopted a peak-hour toll to reduce traffic congestion. The drivers use an electronic card to pay the toll; the card is read by a card reader and the price is deducted from their balance. In addition, cameras record vehicle information and are used for enforcement. Another technique used was that of variable pricing, which was applied to some freeways to charge tolls based on demand at different times of the day. One application of congestion pricing, which was more recent (circa 2006), was the implementation of regional (or cordon) pricing in Stockholm, Sweden. In this implementation, certain areas of the city are designated restricted areas and charge a toll for entry or use of those streets. This resulted in a significant drop (21%+) in traffic in those areas, a corresponding reduction in accidents, and a shift in transport mode from single-occupant (driver) vehicles to mass transit and taxis.

Germany has also been a major location for the implementation of congestion pricing techniques. One such implementation involved truck tolling using global positioning devices (GPS). Trucks and commercial carriers were refitted with GPS, which was used to determine their location in any region. These devices were also able to determine the corresponding toll charges for that location and deduct the appropriate toll from their toll account. Although this GPS/toll system is expensive when compared with the basic transponder in use in most congestion pricing schemes, it removes the need for a toll collection facility; thus the system cost is reduced.

1.2 CONGESTION PRICING IN THE UNITED STATES

Since the early 1990s congestion pricing has been implemented in the United States.

1.2.1 SR-91, San Diego, California

One of the most widely known implementations is that of State Road-91 in San Diego, California.

In this implementation State Road-91 is divided into two separate sections using flexible vertical plastic strips (pylons): One with tolled lanes and another with non-tolled lanes. The tolled lanes charge the drivers a fixed charge for the privilege of using these express lanes. Based on the demand observed, these charges are revised periodically every three months.

1.2.2 I-95, Miami, Florida

Another example of congestion pricing occurs on the I-95 corridor in Miami-Dade County. In this scheme, users are charged a variable fee to drive in the I-95 Express Lanes between the I-395 and the Golden Glades Interchange. The goal of this project is to maintain a speed of 45 mph in the express lanes, and the fee will increase as demand increases to maintain this speed. Buses and high-occupancy vehicles with three or more passengers are allowed to use the Express Lanes for free.

1.2.3 Atlanta, Georgia

Similar tolling schemes are planned or instituted on highways in Georgia. Some studies, which are ongoing in this region, include:

I-85 High-Occupancy Toll Lanes. These lanes will run from Doraville to Gwinnett County, Georgia. Motorcycles and emergency vehicles will be exempt from the toll.

I-75 High-Occupancy Toll Lanes. The prices in this region will dynamically change based on the demand.

I-20 Current High-Occupancy Vehicle Lanes will become High-Occupancy Toll Lanes. Prices will vary with demand.

1.2.4 District of Columbia

In Washington D.C., the study was completed on variable pricing along the I-270. This study examined the feasibility of using Express (Toll) Lanes in place of or in conjunction with the current High-Occupancy Vehicle (HOV) lanes.

1.3 CONGESTION PRICING STRATEGIES

As mentioned earlier, congestion pricing can be implemented in several ways and with a variety of different objectives. In most cases congestion pricing is implemented using the following four techniques:

(1) **HOT Lanes**. Vehicles must have a minimum number of occupants to use the lane. Otherwise, they are charged a toll.

(2) **Express Lanes**. All vehicles are charged a toll to use these lanes during peak periods.

(3) **Cordon Pricing**. An entire region, such as the downtown area, is tolled. All vehicles entering this area must pay a predetermined toll to drive on these streets during certain hours.

(4) **Area Wide Charges**. These are tolls based on mileage (per-mile) driven in a certain area.

When implementing congestion pricing, the method to choose depends on the goal of the authorities involved. Thus, a thorough study must be done to determine the effect of any proposed congestion pricing method on the section of roadway in terms of improved efficiency, throughput, or reduction in dangerous accidents. The choice of the amount of toll to set will be a significant factor in the amount of the reduction in traffic. If the toll is set too low, there will be little or no reduction in traffic, and congestion will remain. Another factor to consider is whether there are enough alternate routes for drivers to switch to when the toll is implemented during peak periods.

Congestion pricing methodologies can be applied to many other industries such as electricity demand and telecommunications demand. Congestion pricing models have been developed for highway toll collection, airline revenue management, and shipping access to ports/docks. In addition, congestion pricing models can be formulated with different objectives, such as minimizing travel time or minimizing system optimal travel time. They may also be formulated to maximize toll revenue, maximize throughput, or maintain a constant speed given increasing demand. As these cases are all different, we see that there may be a variety of objective functions and constraints for the congestion pricing problem, some of which may be nonlinear mathematical programming problems.

1.4 BOOK ORGANIZATION

In this research we will perform a preliminary review based on the different categories of the congestion pricing problem. We will perform the literature review in Chapter 2 based on these different model formulations and solution methods and review three specific implementations of congestion pricing (HOT lanes, Express Lanes) across the United States. The three implementations that are included in the literature review are

(1) State Road 91 (91 Express) San Diego California
(2) Katy Freeway (I-10) in Houston Texas
(3) 95 Express HOT lanes in Miami, Florida

These specific implementations are reviewed based on official reports on studies done for the USDOT and their findings regarding specific performance measures. The 95 Express HOT lanes report by the team at the University of Florida was reviewed in-depth since this is the location of interest for this study. By performing a systematic review, we can show what areas are unexplored and where we can make the greatest contribution to the congestion pricing research.

Chapter 3 introduces current congestion pricing methodologies and discusses the concepts of discrete choice models, utility, and nonlinear programming. Section 3.5 deals with large-scale, nonlinear optimization problems and outlines seven algorithms that have been utilized to solve this type of problem.

In Chapter 4, we formulate the problem and apply the model to I-95 Express to obtain preliminary results. Sample calculations and determination of optimal toll prices are also discussed.

In Chapter 5, we discuss travel demand forecasting and introduce a new method called Generalized Auto-Regressive Conditional Heteroskedasticity (GARCH), which may be applied to real-time data.

Chapter 6 provides a summary and outlines the research conclusions and discusses the results of the analysis tasks and sub-tasks involved in this study.

1.5 SUMMARY

In this introduction we reviewed some implementations, problem formulations, and solution methodologies that are widely applied to the congestion pricing problem. We have seen that over the past 15–20 years several international and domestic regional authorities have implemented congestion pricing to reduce congestion and accidents. We discussed four types of congestion pricing schemes, including HOT lanes, Express Lanes, Area-Wide (mileage) pricing, and Cordon Pricing schema.

Overall, most implementations of congestion pricing have been reportedly successful. However, much more can be done to improve what is known about congestion pricing schemes.

Currently, the USDOT uses software tools for decision-making for congestion pricing. The main tools are Policy Options Evaluation Tool for Managed Lanes (POET-ML), Tool for Rush Hour User Charge Evaluation (TRUCE 3.0), and TRUCE-ST. None of these tools incorporates subsidies for low-income drivers or models the behavior of the system when funds are transferred from users of the HOT lanes to users of the free lanes (such as in the FAIR system adopted in New York). This will be another possible area for future research into optimization pricing.

Other problem approaches may be taken other than those previously mentioned. The Federal Highway Administration (FHWA) of the US Department of Transportation embarked on implementation of congestion pricing projects in four major metropolitan areas (see www.ops.fhwa.dot.gov/publications/fhwahop11030/cm_primer_cs.htm *(retrieved 05/12/13)*)

1. Dallas Ft Worth
2. Puget Sound
3. Minneapolis, St. Paul
4. San Francisco Bay Area

In all these study implementations, the FHWA adopted a four-step process model to help forecast travel demand for use with their congestion pricing scheme. This four-step model includes (1) Travel Demand (Trip Generation), (2) Trip Distribution, (3) Mode choice, and (4) Route Choice. The FHWA found that this four-step model alone is insufficient to develop an effective congestion pricing scheme in any of these four implementations and has refined these steps to include additional information. Some changes included using data to improve the sensitivity of the model to traffic rerouting.

Travel surveys and household income data were also applied to improve the basic four-step model. The FHWA also conducted before and after studies to obtain accurate results for mode choice and travel demand forecasts.

Literature Review

2

2.1 INTRODUCTION

Congestion pricing models have been studied for several years. However, congestion pricing technology and electronic tolling mechanisms such as Auto ID, RFID, and digital camera enforcement technology were not widely implemented until approximately 20 years ago. With the implementation of these technologies came the concept of "Managed lanes" and the analytic studies to evaluate the effectiveness of these pricing schemes.

Ukkusuri et al. performed a comparative analysis of congestion pricing technologies and their implementations. The authors identified different performance criteria for evaluation and utilized these criteria in a formal evaluation framework (ELECTRE IV) algorithm to rank different technologies. The basic methodology of the ELECTRE IV algorithm is divided into three main sections: (1) construction of strong and weak outranking relations, (2) construction of downward and upward ranks, and (3) determination of final ranks.

Some of the congestion pricing technologies used in their evaluation include Radio Frequency ID (RFID), Manual Toll Booths, Automatic Number Plate Recognition (ANPR), Dedicated Short-Range Communication, GPS, and Infrared Communications (IR). The authors conclude that using their performance criteria, RFID ranked first in the list of the six technologies tested with GPS, Infrared tied for third place, and Manual Toll Booth (MTB) ranked last in terms of performance with the ELECTRE IV algorithm.

Labi and Issariyanukula (2011) examine the exploitation of real-time technology for use in congestion pricing (CP) implementation. Their NEXTRANS report discusses the financial and technical feasibility of Congestion Pricing as a revenue generation source in Indiana.

DOI: 10.1201/9781003358701-2

The authors compared both static and dynamic congestion pricing techniques with the base case of building a new untolled free lane. The results of their study demonstrated that with static pricing efficiency decreases, whereas with dynamic congestion pricing efficiency and throughput increase.

Dynamic pricing provides a reduction in overall user cost, an increase in traffic volume, and reduction in the duration of the peak period, as compared to the base case of two free lanes.

Modi et al. investigated the implementation of dynamic congestion pricing along the I-95 Express in Miami, FL. Their report answered several important questions, including whether motorists shift their travel times in anticipation of toll volatility during the day. The authors compared the dynamic pricing (DP) method to time-of day (TOD) pricing to determine which method performs better in the HOT lanes along the 95 Express. The study examined how the number and placement of entry points and exit points affect the operations of the express lanes along I-95. The tolling algorithms implemented for 95 Express were evaluated, as were other possible enhanced algorithms that are applicable to the express lanes.

The I-95 Express algorithm, as presented by the authors, is based on a look-up table that uses traffic densities as well as changes in measured traffic densities to determine an optimal toll price to meet the objectives of the I-95 Express authorities. One of these objectives is to maintain free-flow conditions along the express lanes, which is considered to be a speed of 45 mph or above.

The I-95 express lanes are monitored by 31 loop detectors that provide real-time information on speed, density, and traffic volume to an automated system and are displayed to an operator. This information is updated every 15 minutes, and the toll is adjusted accordingly at these intervals. The toll price varies from as low as $0.25 to as high as $7.25 depending on traffic conditions.

The authors of the University of Florida study present the I-95 Express tolling algorithm and toll look-up table, which are based on level of service (LOS) from A-F, traffic density, and change in traffic density as parameters in the final toll determination. Using a cell transmission macroscopic model (Daganzo, 1993) the authors analyze travel demand and express lane user behaviors as they relate to departure times and the volatility of toll rates.

A capacity analysis was performed on the effect of delineators used for the 95 express lanes, and the results show that the capacity of the general purpose (GP) lane increased after delineators were set up and tolling started in the express lanes, while the capacity of GP lane 1 in the vicinity of entry/exit points reduced after delineators were installed and tolling started

Similarly, results for the effect of delineators on the HOT lanes were determined. It was found that the capacity of the HOT lanes reduced after tolling started. The speed of HOT lanes varies depending on the vicinity of entry/exit points.

The study presents the optimization objective function procedure for the I-95 express lanes:

$$\text{Maximize S} = \sum GP\_speed_i + M * min \left[EL\_Speed_i - 45,0 \right] \qquad (1)$$

where $GP\_speed(i)$ is the speed on the general-purpose lanes at interval I, $EL\_speed(i)$ is the speed on the express lanes at interval I, and M is a penalty parameter. For this objective, M is set to be equal to 100.

2.1.1 Dynamic Pricing Algorithm of I-95 Express (Phase 1)

The I-95 express lanes utilize a look-up table based on traffic density during each time (i) and the change in traffic density between time (i) to time $(i+1)$. When the magnitude of the change in traffic density is sufficient, the toll will be increased or decreased according to the pre-determined values in the table. The algorithm is presented here as outlined in the University of Florida study (Modi et al.):

(1) Calculate average traffic density of HOT lane $D(t)$. $D(t)$ may be corrected for factors such as entry/exit point or weaving.
(2) Calculate the change in density $\Delta D = D(t) - D(t-1)$, where $D(t)$ is the traffic density at time t and $D(t-1)$ is the density at time t − 1.
(3) Determine the toll adjustment amount ΔR from the Delta Setting Table (look-up table)
(4) Calculate the new toll amount as $R(t) = R(t-1) + \Delta R$
(5) Compare the new toll amount with the minimum and maximum toll amounts in the LOS table. If the new toll amount is greater than the maximum value, then use the maximum toll in the table. If the new toll amount is less than the minimum value, then use the minimum value.

The DTS look-up table (Table 2.1) as presented in the University of Florida Research study is reproduced here for convenience. This look-up table represents the optimal solution to the maximization problem formulation of the congestion pricing problem along I-95 Express. The sub-ranges in the LOS categories, i.e., 0–11 in LOS A, are based on traffic conditions for this route. Also, the change in densities −1 to −6 and +1 to +6 is based on experience in the delta jumps.

TABLE 2.1   Express DTS Look-Up Table (Negative Density)

LOS	TRAFFIC DENSITY (VPMPL)	CHANGE IN TRAFFIC DENSITY					
		-6	-5	-4	-3	-2	-1
	0	-$0.25	-$0.25	-$0.25	-$0.25	-$0.25	-$0.25
	1	-$0.25	-$0.25	-$0.25	-$0.25	-$0.25	-$0.25
A	2	-$0.25	-$0.25	-$0.25	-$0.25	-$0.25	-$0.25
	3	-$0.25	-$0.25	-$0.25	-$0.25	-$0.25	-$0.25
	4	-$0.25	-$0.25	-$0.25	-$0.25	-$0.25	-$0.25
	5	-$0.25	-$0.25	-$0.25	-$0.25	-$0.25	-$0.25
	6	-$0.25	-$0.25	-$0.25	-$0.25	-$0.25	-$0.25
	7	-$0.25	-$0.25	-$0.25	-$0.25	-$0.25	-$0.25
	8	-$0.25	-$0.25	-$0.25	-$0.25	-$0.25	-$0.25
	9	-$0.25	-$0.25	-$0.25	-$0.25	-$0.25	-$0.25
	10	-$0.25	-$0.25	-$0.25	-$0.25	-$0.25	-$0.25
	11	-$0.25	-$0.25	-$0.25	-$0.25	-$0.25	-$0.25

LOS	TRAFFIC	CHANGE IN TRAFFIC DENSITY					
	12	-$0.50	-$0.50	-$0.50	-$0.25	-$0.25	-$0.25
B	13	-$0.50	-$0.50	-$0.50	-$0.25	-$0.25	-$0.25
	14	-$0.50	-$0.50	-$0.50	-$0.25	-$0.25	-$0.25
	15	-$0.50	-$0.50	-$0.50	-$0.50	-$0.25	-$0.25
	16	-$0.50	-$0.50	-$0.50	-$0.50	-$0.25	-$0.25
	17	-$1.25	-$1.00	-$0.75	-$0.50	-$0.25	-$0.25
	18	-$1.25	-$1.00	-$0.75	-$0.50	-$0.25	-$0.25
	19	-$1.25	-$1.00	-$0.75	-$0.50	-$0.25	-$0.25
	20	-$1.25	-$1.00	-$0.75	-$0.50	-$0.25	-$0.25
C	21	-$1.25	-$1.00	-$0.75	-$0.50	-$0.25	-$0.25
	22	-$1.25	-$1.00	-$0.75	-$0.50	-$0.25	-$0.25
	23	-$1.25	-$1.00	-$0.75	-$0.50	-$0.25	-$0.25
	24	-$1.25	-$1.00	-$0.75	-$0.50	-$0.25	-$0.25

LOS	TRAFFIC	CHANGE IN TRAFFIC DENSITY					
	25	-$1.25	-$1.00	-$0.75	-$0.50	-$0.25	-$0.25
	26	-$1.25	-$1.00	-$0.75	-$0.50	-$0.25	-$0.25
	27	-$1.50	-$1.25	-$1.00	-$0.75	-$0.50	-$0.25
	28	-$1.50	-$1.25	-$1.00	-$0.75	-$0.50	-$0.25
	29	-$1.50	-$1.25	-$1.00	-$0.75	-$0.50	-$0.25

LOS	TRAFFIC	CHANGE IN TRAFFIC DENSITY					
D	30	−$1.50	−$1.25	−$1.00	−$0.75	−$0.50	−$0.25
	31	−$1.50	−$1.25	−$1.00	−$0.75	−$0.50	−$0.25
	32	−$1.50	−$1.25	−$1.00	−$0.75	−$0.50	−$0.25
	33	−$1.50	−$1.25	−$1.00	−$0.75	−$0.50	−$0.25
	34	−$1.50	−$1.25	−$1.00	−$0.75	−$0.50	−$0.25
	35	−$1.50	−$1.25	−$1.00	−$0.75	−$0.50	−$0.25
	36	−$1.50	−$1.25	−$1.00	−$0.75	−$0.50	−$0.25
	37	−$1.50	−$1.25	−$1.00	−$0.75	−$0.50	−$0.25

LOS	TRAFFIC	CHANGE IN TRAFFIC DENSITY					
E							5
	38	−$1.50	−$1.25	−$1.00	−$0.75	−$0.50	−$0.25
	39	−$1.50	−$1.25	−$1.00	−$0.75	−$0.50	−$0.25
	40	−$1.50	−$1.25	−$1.00	−$0.75	−$0.50	−$0.25
	41	−$1.50	−$1.25	−$1.00	−$0.75	−$0.50	−$0.25
	42	−$1.50	−$1.25	−$1.00	−$0.75	−$0.50	−$0.25
	43	−$1.50	−$1.25	−$1.00	−$0.75	−$0.50	−$0.2 5
	44	−$1.50	−$1.25	−$1.00	−$0.75	−$0.50	−$0.2 5
	45	−$1.50	−$1.25	−$1.00	−$0.75	−$0.50	−$0.25
F	>45	−$2.00	−$2.00	−$2.00	−$2.00	−$1.00	−$0.25

(*Source:* Modi et al.)

TABLE 2.2 Express DTS Look-Up Table (Positive Density Shift)

LOS	TRAFFIC DENSITY (VPMPL)	CHANGE IN TRAFFIC DENSITY					
		1	2	3	4	5	6
A	0	$0.25	$0.25	$0.25	$0.25	$0.25	$0.25
	1	$0.25	$0.25	$0.25	$0.25	$0.25	$0.25
	2	$0.25	$0.25	$0.25	$0.25	$0.25	$0.25
	3	$0.25	$0.25	$0.25	$0.25	$0.25	$0.25
	4	$0.25	$0.25	$0.25	$0.25	$0.25	$0.25
	5	$0.25	$0.25	$0.25	$0.25	$0.25	$0.25
	6	$0.25	$0.25	$0.25	$0.25	$0.25	$0.25
	7	$0.25	$0.25	$0.25	$0.25	$0.25	$0.25
	8	$0.25	$0.25	$0.25	$0.25	$0.25	$0.25
	9	$0.25	$0.25	$0.25	$0.25	$0.25	$0.25
	10	$0.25	$0.25	$0.25	$0.25	$0.25	$0.25
	11	$0.25	$0.25	$0.25	$0.25	$0.25	$0.25

(*Continued*)

TABLE 2.2 (Continued)

LOS	TRAFFIC DENSITY (VPMPL)	CHANGE IN TRAFFIC DENSITY					
		1	2	3	4	5	6
B	12	$0.25	$0.25	$0.25	$0.50	$0.50	$0.5 0
	13	$0.25	$0.25	$0.25	$0.50	$0.50	$0.5 0
	14	$0.25	$0.25	$0.25	$0.50	$0.50	$0.5 0
	15	$0.25	$0.25	$0.50	$0.50	$0.50	$0.5 0
	16	$0.25	$0.25	$0.50	$0.50	$0.50	$0.50
	17	$0.25	$0.25	$0.50	$0.75	$1.00	$1.2

(*Source:* Modi et al.)

LOS	TRAFFIC	CHANGE IN TRAFFIC DENSITY					
						5	
	18	$0.25	$0.25	$0.50	$0.75	$1.00	$1.25
C	19	$0.25	$0.25	$0.50	$0.75	$1.00	$1.25
	20	$0.25	$0.25	$0.50	$0.75	$1.00	$1.25
	21	$0.25	$0.25	$0.50	$0.75	$1.00	$1.25
	22	$0.25	$0.25	$0.50	$0.75	$1.00	$1.25
	23	$0.25	$0.25	$0.50	$0.75	$1.00	$1.25
	24	$0.25	$0.25	$0.50	$0.75	$1.00	$1.25
	25	$0.25	$0.25	$0.50	$0.75	$1.00	$1.25
	26	$0.25	$0.25	$0.50	$0.75	$1.00	$1.25
	27	$0.25	$0.50	$0.75	$1.00	$1.25	$1.50
	28	$0.25	$0.50	$0.75	$1.00	$1.25	$1.50
	29	$0.25	$0.50	$0.75	$1.00	$1.25	$1.50
	30	$0.25	$0.50	$0.75	$1.00	$1.25	$1.50
D	31	$0.25	$0.50	$0.75	$1.00	$1.25	$1.50
	32	$0.25	$0.50	$0.75	$1.00	$1.25	$1.50
	33	$0.25	$0.50	$0.75	$1.00	$1.25	$1.50
	34	$0.25	$0.50	$0.75	$1.00	$1.25	$1.50
	35	$0.25	$0.50	$0.75	$1.00	$1.25	$1.50

LOS	TRAFFIC	CHANGE IN TRAFFIC DENSITY					
	36	$0.25	$0.50	$0.75	$1.00	$1.25	$1.50
E	37	$0.25	$0.50	$0.75	$1.00	$1.25	$1.50
	38	$0.25	$0.50	$0.75	$1.00	$1.25	$1.50

LOS	TRAFFIC	CHANGE IN TRAFFIC DENSITY					
	39	$0.25	$0.50	$0.75	$1.00	$1.25	$1.50
	40	$0.25	$0.50	$0.75	$1.00	$1.25	$1.50
	41	$0.25	$0.50	$0.75	$1.00	$1.25	$1.50
	42	$0.25	$0.50	$0.75	$1.00	$1.25	$1.50
	43	$0.25	$0.50	$0.75	$1.00	$1.25	$1.50
	44	$0.25	$0.50	$0.75	$1.00	$1.25	$1.50
	45	$0.25	$0.50	$0.75	$1.00	$1.25	$1.50
F	>45	$0.50	$1.00	$2.00	$2.00	$2.00	$2.00

The researchers at the University of Florida (Modi et al.) describe a method for solving their optimization problem called a "GA procedure." In this method a large-scale optimization problem is solved by utilizing a pool of individual solutions ("parents"), which are then randomly selected to generate new solutions called "offspring," and the process repeats itself iteratively until the "optimal" solution is reached.

The GA procedure as described in the University of Florida study (Modi et al.), as represented in Figure 2.1.

As outlined in the 95 Express report (Modi et al., 2012), the GA procedure is initialized with ten randomly selected "individuals" to start the iteration.

These individuals are each a non-optimal solution to the problem, which is then evaluated by the algorithm for fitness. The solution with the best fitness value is then selected to continue the iteration. Each step in the GA process is then followed until 100 iterations are completed. After 100 iterations, the best solution obtained is then considered the "optimal" solution to the problem.

Carson (2005) outlines six steps for a successful monitoring and evaluation program for managed lane facilities:

(1) Setting objectives for the system that reflect the programs desired performance
(2) Identifying performance measures to evaluate goals and objectives
(3) Identifying data sources to utilize in the calculation of performance measures
(4) Defining proper evaluation methods for the data
(5) Scheduling the periodic monitoring of the system
(6) Reporting results in an easily understood format

In the report, Carson specifically addresses the difference between managed lane operations and general freeway operations and develops best practices for managed lanes. The paper documented specific managed lane benefits, which may be used to develop benchmarks for managed lane monitoring.

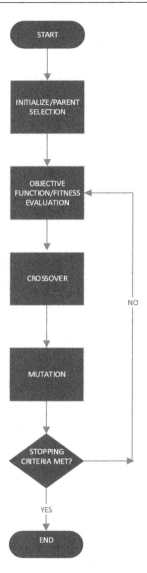

FIGURE 2.1 GA Algorithm Flowchart

(Source: Modi et al.)

The report described three factors that are influential in the monitoring of managed lane performance:

(1) Accessibility, including number of entry and egress points
(2) Hours of operation of the facility
(3) Eligibility criteria, including toll rates, vehicle types and occupancies, etc.

The author listed four methods for providing access to managed lanes. These are (1) Direct Merges, (2) Slip Ramps, (3) Direct Access Ramps, and (4) Direct Connections from other lanes. Seven different goals and objectives for managed lane facilities were outlined by the author in this report. They are:

(1) MOBILITY/CONGESTION GOAL: Increase mobility during recurring and non-recurring congestion
OBJECTIVES: Increase speed, increase throughput, decrease travel times, decrease delay
(2) RELIABILITY GOAL: Increase reliability during recurring and nonrecurring congestion
OBJECTIVES: Decrease travel speed or travel time variation. Increase "on-time" performance
(3) ACCESSIBILITY GOALS: Increase accessibility while reducing congestion OBJECTIVES: Maintain or increase lane miles along the facility and decrease the number of facility restrictions
(4) SAFETY GOALS: Increase safety
OBJECTIVES: Decrease frequency or severity of accidents and decrease incident duration
(5) ENVIRONMENTAL GOALS: Decrease impacts to the environment
OBJECTIVES: Decrease air quality pollutants and decrease noise pollution.
(6) SYSTEM PRESERVATION GOAL: Increase system service life
OBJECTIVES: Decrease system deficiency.
(7) ORGANIZATIONAL EFFICIENCY GOALS: Increase productivity
OBJECTIVES: Maximize revenue, minimize costs, and increase system performance

Carson (2005) provides a set of recommended performance measures for evaluating ITS in the study. The recommendations include guidelines for safety, throughput and capacity, productivity, energy and environment, customer satisfaction and mobility.

Table 2.3 documents the recommended measures.

The study concluded with a review of different types of managed lane facilities and notes that despite differences in the types of facilities, commonalities exist for the development of monitoring and performance evaluation. The report notes that, generally, passenger-focused managed lane facilities have an interest in increasing throughput as measured by average vehicle occupancy (AVO) levels combined with an increase in vehicle travel speeds.

The author specifically points out that HOT lanes "present unique opportunities for toll revenue," and other factors (such as environmental) may be placed in secondary interest in these facilities. A summary of the performance measures across facility types, including HOV, HOT, Exclusive Lanes (Passenger/Freight), mixed flow separation (Passenger/Freight), Lane Restrictions (Freight) as well as dual facilities (Passenger/Freight), concluded the investigation.

Cambridge Systematics developed and presented an Evaluation Plan Framework for I-95 Express managed lanes (2009). In the report, they combined performance measures into three groups. Group 1 includes corridor performance and utilization, Group 2 includes operation and efficiency, and Group 3 includes customer satisfaction

Group 1 performance measures are listed as:

1. Traffic-Express Lanes versus GP Lanes
2. Transit-Express Bus Rapid Transit (BRT)
3. Other

TABLE 2.3   Recommended Dynamic Pricing Performance (Carson, 2005)

GOAL AREA	PERFORMANCE MEASURE
Safety	Reduction in overall crash rate
	Reduction in rate of fatal crashes
	Reduction in rate of injury crashes
	Improve surrogate measures
Mobility	Reduction in travel time delay
	Reduction in travel time variability
	Improvement in surrogate measures
Throughput/capacity	Increase in throughput or capacity
Customer satisfaction	Difference between users' expectation and experience in relation to product
Productivity	Cost savings
Energy and environment	Reduction in emissions, reduction in fuel consumption

Group 2 performance measures are listed as:

1. Operational Efficiency-Express Lanes and GP Lanes
2. Operational Efficiency-Bus and Rapid Transit Express

Group 3 performance measures are listed as:

1. Acceptance/Customer Satisfaction of Express Lanes
2. Acceptance/Customer Satisfaction level of Bust Rapid Transit

For each evaluation measure, the performance metrics, the Miami/FDOT objective, and USDOT/UPA objective were listed.

As an example, for the Traffic (EL versus GP lanes) the performance metrics were listed as traffic volume, average speed, travel time saved, LOS, AVO, and vehicle and person throughput. For the Miami/FDOT objective, ML (express lane) optimization, congestion relief in Express Lane, congestion relief in GP lane, maintain free flow, and express bus (BRT) were included. The USDOT/UPA objectives listed were congestion, tolling, movement of goods, and transit.

Vaze and Barnhart (2012) investigate congestion pricing in the airline industry. The paper documented a model of airline frequency in the environment of congestion pricing. Competition between airlines is viewed as a measure of the willingness of airlines to pay for airport slots. The paper evaluated the impact of congestion prices on the different stakeholders and investigated how the efficiency of congestion pricing schemes impact frequency competition in particular markets. The authors note that most models in the literature assume constant load factors and constant aircraft sizes, and this does not account for the significant factor of variability in the number of passengers per flight, which plays a vital role in the effectiveness of congestion pricing for airlines. The paper found that marginal cost pricing is more effective than flat pricing in reducing congestion without penalizing airlines exceedingly.

Tuerk et al. (2012) studied the effect of London congestion charges on traffic volumes when the toll is increased from £5 to £8 in central London and the implementation of the Western extension zone.

The methodology employed by the authors involves difference-in-differences estimation to determine estimates of the effects of these interventions. The results of this study imply an 8% reduction in traffic due to the increase in congestion price and a 10% reduction in traffic due to the implementation of the Western extension zone.

Franklin (2012) examines the role of contextual variables in the equity effects of congestion pricing. The potential for inequity effects of congestion pricing has come into focus as a possible hindrance for its implementation as a demand management strategy around the world. But many argue that contextual factors such as automobile access, work schedule, flexibility, and spatial distribution of activities are independent variables in determining how the burden of the toll should fall on different sectors of the demographics. The paper uses structural equation modeling to apply to Stockholm's congestion (CP) implementation to estimate the roles of age, gender, and income as independent variables. The finding of the paper indicates that only gender was a significant factor in total effects. But contextual factors such as access to car, possession of transit pass, and a workplace on the same side of the Stockholm cordon were significant mediating effects on the demographic variable's effects on trips by automobile.

Zheng et al. (2012) studied the impacts of combining a macroscopic model of congestion network with an agent-based simulator to study dynamic cordon congestion pricing schemes. The agent-based simulator can represent complex travel behavior relating to departure time choice and heterogeneous users.

The authors argued that traditional traffic simulators consider traffic demand as inelastic to level of congestion, whereas most congestion pricing models are sensitive to demand fluctuations and non-stationary conditions. In addition, most pricing models assume a deterministic homogeneous population.

The paper showed how the output of a multi-agent-based simulator is consistent with the physics of traffic flow dynamics. They then develop a dynamic cordon congestion pricing scheme, and their results showed that the travel time savings at both the aggregated and dis-aggregated level outweigh the costs. The traffic congestion within the cordon is reduced without being shifted to outside the cordon. The paper concluded that equity issues can be further investigated if provided with more information about the income of agents.

Ohazulike et al. (2012) performed a multi-objective congestion pricing analysis using a game-theoretic approach in this paper.

In this study, the authors developed a global optimization problem formulation for the congestion pricing (problem. It included objectives for traffic congestion, air pollution, noise, and safety as conflicting objectives in solving the problems of optimal road pricing. Using a game-theoretic approach and the concept of Nash Equilibrium (NE) the authors extended the single objective (Stackelberg game) to that of multi-authority game with conflicting objectives. They developed a road pricing scheme for this multi-objective problem formulation and proved that no pure NE in general exists under these conditions. However, NE may be possible in specific instances. The authors designed a system that produces a pure NE while simultaneously inducing cooperative behavior among participants, thus resulting in optimal tolls.

Yang et al. (2012) formulated the pricing problem as a stochastic macroscopic traffic flow model and developed the methodology to find a distance-based dynamic pricing strategy for managed toll lanes to maximize revenue. The authors proposed a simulation-based algorithm to obtain optimal pricing in real time.

The authors developed a partial differential equation for the traffic evolution in the GP lane. They claim that the dynamic distance-based algorithm developed in this paper can also be applied to other objective functions such as maximizing throughput. The authors also inferred that their strategy will be applicable to the classical LWR model.

Zhong et al. (2012) presented a paper titled "A Reliability-Based Stochastic System Optimum Congestion Pricing Model Under ATIS With Endogenous Market Penetration and Compliance Rate."

In their paper the authors divided all travelers into two classes. In the first class, travelers who follow ATIS advice are considered to be "equipped." The other class of travelers is unaware or does not follow ATIS advice and is considered to be "unequipped" (or non-compliant). Travelers consider pricing (CP), travel times, and network reliability when making travel decisions about route choices. The ATIS market penetration, according to the authors, can be measured as an increasing function of the information benefit, and ATIS compliance rates are given by the probability of the travel costs of equipped travelers being less than or equal to that for unguided drivers. The authors formulated the model as an equivalent variation inequality problem, assuming that the origin–destination (OD) pair travel demand is found from the minimal perceived travel cost between the (OD) pair.

Morgul and Ozbay (2011) investigated a simulation evaluation of a feedback-based dynamic congestion pricing strategy on alternate locations.

The paper proposed a methodology for extending the concept of dynamic tolling to two neighboring tolled facilities. The authors argue that current tolling algorithms can result in highly fluctuating toll prices that vary widely and change frequently over short time intervals, which can confuse and inconvenience travelers.

The authors addressed the problem by proposing a less reactive tolling algorithm, which is applied to two parallel routes in the New York/New Jersey metro area. The algorithm was tested on two tunnels between New York and New Jersey using a microscopic simulation of the traffic entering New York City.

Yang and Chu (2011) developed a stochastic model for traffic flow prediction and its validation. The authors also stated that such a model has applicability to the congestion pricing problem. They argued that traditional deterministic traffic flow models such as the LWR model do not capture the full scope of factors such as erratic driver behavior or weather and need to be replaced with the stochastic model.

The proposed model uses partial stochastic differential equations to predict traffic evolution along the highway corridor. The authors calibrated and validated their model using real data and claimed a higher power to predict traffic behavior than a deterministic model.

Ozbay et al. (2011) investigated a mesoscopic simulation evaluation for dynamic congestion strategies for New York City crossings. In their study the authors performed a simulation-based evaluation of dynamic pricing at these crossings using a mesoscopic simulation model with a stepwise tolling algorithm. The authors presented an estimation of the value of time (VOT) for different classes of travelers. In particular, commercial and commuter traffic is studied to determine the VOT for tolling purposes. VOT for New York City region commercial vehicles is derived using a logit model of stated preference (SP) data.

The authors conclude the paper with an analysis of simulated dynamic prices as compared with static pricing methods for this region.

Michalaka et al.'s (2011) study titled "Proactive and Robust Dynamic Pricing Strategies for High-Occupancy Toll Lanes" developed a new, robust system methodology for dynamic pricing of toll lanes. The authors tested their algorithm on 95 Express in Miami, Florida. They observed that most studies in the literature are based on hypothetical situations and do not address the uncertainty involved in demand when calculating toll rates. The authors claimed to have developed a robust approach to dynamic pricing toll rates along I-95 based on real-time conditions.

The approach consists of a scenario-based robust toll optimization. Simulation is conducted along I-95 to demonstrate the new approach.

In their paper the authors differentiated between the term "congestion pricing" and the narrower term of "dynamic pricing," which applies to time-varying toll rates based on measured demand in the system.

Wadoo et al. (2011) presented a feedback-based dynamic congestion pricing model with alternate routes available. The model utilizes queuing theory and traffic conservation laws for its derivation. It also uses fundamental macroscopic modeling relationships in its development.

The authors used a logit model for the pricing and driver choice behavior relationship. They used this model to derive a feedback control law that uses real-time information to come up with the toll price.

The paper then demonstrated the law using a simulation of the derived feedback control model.

Klodzinski and Adler (2010) worked on the development of travel demand forecasts for estimating express lane traffic and variable toll rates.

The authors argued that most existing regional traffic forecasting models do not provide hour-by-hour traffic volumes.

Since toll calculations depend heavily on hour-by-hour traffic conditions, it is vital to perform additional hourly analysis to provide the needed information to accurately calculate the toll. The paper presented a spreadsheet-based procedure that estimates hourly usage for the express lanes. This hourly forecast estimate is based on the daily forecast or period-specific forecasts by a regional forecasting model.

The Express Lane Time of Day (ELTOD) forecast inputs are total daily traffic volume, geometric configuration of the facility, and tolling policy.

ELTOD solves for supply/demand equilibrium for each hour to estimate the split that occurs between GP and Express Lanes.

Changes in the ELTOD parameters can be used to test a congestion pricing strategy for demand management and a required LOS in the express lanes. The ELTOD procedure was tested using a stretch of Interstate 75 in Southwestern Florida.

Jang and Chung (2010) propose a method for dynamically determining toll prices in response to changes in traffic conditions. The pricing strategies include revenue maximization and delay minimization along the express lanes. They presented methods for estimating the parameters for the model and applied them to a 14-mile stretch along the California Freeway in the San Francisco Bay Area.

The study concluded that utilizing all the available unused HOV lane capacity for HOT lanes does not maximize the revenue. Only a certain percent of the unused portion of the available should be used to convert into HOT lanes to achieve maximum revenue.

Gardner et al. (2010) examine the development of congestion pricing for transportation networks with uncertain supply and demand. The authors were motivated to find what role uncertainty in supply and demand information plays in transportation networks where tolling is practiced.

The authors attempted to quantify the effect of providing information to travelers with/without tolling when there exists uncertainty in both supply and demand. The paper used a scenario-based solution methodology where both users and the operators have flexibility to make travel choices based on the available information about supply and demand. The solution techniques were applied to the Sioux Falls transportation network.

Zhong (2009) examined dynamic congestion pricing for multi-class, multi-mode transportation systems with asymmetric cost functions.

The focus of the paper was to look into dynamic congestion pricing (CP) to determine optimal time-varying tolls for a queuing network.

The authors utilize a combined space-time expanded network and conventional equilibrium modeling technique to develop a multi-class, multi-mode, and multi-criteria traffic equilibrium model. The symmetric cost function model is extended to deal with the interactions between buses and cars. The authors concluded that there exists a link toll pattern that can be used to drive a multi-class, multi-mode, multi-criteria user equilibrium flow to system optimum when the system objective function is measured in money.

Lu and Mahmassani (2010) developed a "Dynamic Pricing with Heterogeneous users: Gap Driven Solution Approach for Bicriterion Dynamic User Equilibrium Problem."

In this paper the authors discuss the Bicriterion Dynamic User Equilibrium (BDUE) problem and how it characterizes dynamic user equilibrium in networks representing path choice interactions of users with different VOTs. The BDUE is seen as a better way of accommodating behavioral and policy realism in applying dynamic pricing schemes. The paper attempted to gain path-flow patterns satisfying the BDUE conditions by performing a study to adapt the gap-driven a simulation-based algorithmic framework to solve the DUE problem with a constant VOT. Essentially, the algorithm is:

(i)   A column generation approach that integrates a simulation-based dynamic loading model that captures traffic dynamics and determines travel times for a given path-flow pattern

(ii) A path generation scheme that partitions the entire VOT and determines the multiple user classes and least cost paths for each user class

(iii) A multi-class flow equilibrating method for updating the current path assignment

The authors conclude by presenting results that show that convergence of their proposed algorithm is not affected by different VOT assumptions.

Iseki et al. (2010) examined the links between electronic roadway tolling technologies and road pricing policy objectives.

In this paper the authors argue that there is no "best" tolling technology, but rather the optimal configuration depends on the policy objectives of the tolling effort.

Factors such as facility type and regional scope are major factors in the decision. The paper examined eight road pricing programs. For each program, the authors examined three technical tasks, nine technology sets, and six policy objectives.

The paper found that there are two predominant factors that determine the type of roadway technology adopted:

(1) Geographical scale of road network and
(2) Complexity of fee calculation

The authors conclude that issues facing implementation of road pricing technologies are less about the technologies or the road pricing objectives and more about the economics and political linking of the two.

Brands et al. (2009) utilize a pattern search algorithm to develop optimal toll design in dynamic traffic networks. Such a design problem is formulated as a bi-level mathematical program. The upper level minimizes an objective function, i.e., average travel time in network, and in the lower level a dynamic traffic assignment model is used to determine the effect of differentiated road pricing schemes on traffic. The paper focused on the upper-level optimization problem and tested variants of a pattern search algorithm with a case study. In this study the authors showed that many different local minima exist and that the optimal prices are the same for each one.

The case study tested several variants of the pattern search algorithm and showed that it may be beneficial to change more variables at each iteration to achieve the optimal solution.

DeCorla-Souza (2009) investigated the concept for peak-period pricing along metropolitan freeway systems. The study examined the concept of pricing on all lanes of a freeway system and showed that neither pre-set tolling (Time-of-day) nor dynamic tolling would provide for optimal pricing when all lanes of a freeway are tolled. The paper developed a concept that would allow for such freeway-wide tolling on all lanes. The concept includes providing information to travelers about highway conditions before they leave home, and simultaneously ensuring maximum utilization of the freeway system with a pre-determined LOS set by policy.

Lu and Mahmassani (2008) examine "Modeling user response to pricing: Simultaneous route and departure time Network Equilibrium with Heterogeneous Users."

The authors presented a generalized framework to incorporate route and departure time as well as heterogeneity. They present a multi-criterion simultaneous route and departure time user equilibrium (MSRDUE) along with a simulation-based algorithm. The model explicitly looks into travelers with different values of time (VOT).

The authors formulated the problem as an infinite-dimensional variational inequality problem and solved it as a column-generation-based framework that embeds an alternative-based algorithm that finds the VOT, value of early schedule delay (VOESD), and value of late schedule delay (VOLSD) breakpoints that define multiple user classes and the least trip cost for each user class. A traffic simulator captured the traffic flow dynamics and determined the travel costs experienced. The algorithm also included a path-swapping alternative flow scheme to solve the restricted multi-class SRDUE. The scheme is applied to an actual network that shows the importance of capturing user heterogeneity and time shifts.

Mahmassani et al. (2005) examine "Toll Pricing and Heterogeneous Users: Approximation Algorithms for Finding Bi-criterion Time Dependent Efficient Paths in Large Scale Networks."

This study examines the algorithms for finding exact and approximations for efficient time-dependent shortest paths for use with dynamic traffic assignment applications to variable toll pricing networks. The authors presented a least-cost generalized path algorithm that determines a complete set of efficient time-dependent paths that consider travel time and cost simultaneously.

Due to the computational complexity, exact solutions may not be practical for large networks. Approximation (heuristic) algorithms are devised and tested using the concept of efficiency in multi-objective shortest path problems within a binary search framework to find a set of extreme efficient paths to minimize expected error in the approximation.

Experimental results showed that the computation time and size of the solution set are determined by parameters as the number of nodes in the network and number of time intervals. Test results showed that the approximation scheme is efficient for large-scale bi-objective time-dependent shortest path applications.

Sullivan (2000) evaluated the impacts of the *State Road 91 (SR 91) Ex*press Lane Variable-Toll Facilities.

The SR 91 express lanes opened in 1995 with four tolled lanes in both directions and State Route 91 in Orange County, California. The tolls vary on a fixed schedule and are updated periodically based on monitored traffic conditions.

The study began with approximately a year and a half of observations prior to the opening of the toll road to establish a baseline of traffic conditions and demand, and to be able to compare before and after scenarios with and without the implementation of the toll. The report provided detailed traffic measurements on such parameters as vehicle occupancy counts, transit ridership information, and travel surveys with commuters.

The author performed data analysis on route choice (toll/no toll), AVO levels, transponder acquisition, and also time of day choices for commuters.

In the report, the author notes several significant changes to the SR91 toll road since it opened in 1995:

(1) Change to toll schedule, which allowed for the charge to be changed on an hour-by-hour basis. Previously, the toll was a fixed amount throughout the four-hour peak period on weekdays and the six-hour peak period on Fridays.

(2) Initially, high occupancy vehicles with more than three travelers (HOV3+) could use the toll road for free. In 1998, this rule changed so that HOV3+ users were now charged at a rate of 50% of the published toll.

(3) The opening of an alternate toll road in October 1998 (Eastern Toll Road), which competes with the SR 91 for travel to Irvine County.

(4) The intention of the California Private Transportation Company to sell the business to a non-profit entity.

These changes have impacted the operation of the SR 91 toll facility, and the report attempts to measure and document the resulting effects on the facility.

The report described the SR 91 express lanes as being located between the 91/55 junction in Anaheim, CA, and the Orange County/Riverside line. The two express lanes (in either direction) are situated in the median of the freeway and separated by soft pylons from the GP lanes. Prior to the opening of SR 91 express lanes, delays of 40 minutes were typical, whereas delays were reduced to 10 minutes after the construction of the new free lanes but have returned to higher levels again.

The SR91 Express is a 10-mile-long facility that was constructed as a for-profit private investment. Tolls on the SR 91X vary hour by hour and reflect the travel time savings of those in the toll lanes as compared to those travelers in the free lanes.

Tolls vary from a low of $0.75 to a high of $3.75 during the rush hour.

A "frequent-user club" called the 91X Express club pays a flat fee of $15 per month and receives a $0.75 saving off each tolled trip made. This plan typically results in an overall savings for travelers who use the 91 express lanes a minimum of 20 times per month.

About 360,000 transponders had been issued by the time of the report (1999), and they all utilize a read/write radio frequency (RF) tag technology

The 91 Express report concludes that the toll lanes are a well-managed, well-accepted alternative choice for many travelers who are willing to pay to avoid traffic congestion and enjoy a reliable travel time savings in Southern California. The author noted favorable impacts on the corridor due to greatly improved travel conditions and increase in vehicular traffic.

The 91X corridor has also seen a dramatic increase in HOV3+ traffic since its inception, which helps to improve throughput and reduce air pollution.

The increase in Single Occupancy Traffic was seen to be due to three reasons:

(1) Traffic returning from parallel streets
(2) New travelers who had previously avoided this overly congested roadway
(3) A continuation of the SOV growth trend that had been present before the implementation of the 91 express lanes

In performing this study, the author adopted a methodology to analyze the data from the 91Xpress and extract useful information.

The study methodology included observations of traffic conditions along the 91X freeway and some control sites. Information recorded included speeds, vehicle types, traffic counts, and vehicle occupancies.

Also included are observations of volumes along specific ramps and speed on parallel arterials. These values are used to estimate the amount of traffic diversion during peak-hour toll periods. The third type of observations included ridership on public transportation on ride-share programs.

Travel survey data was collected to understand the traveler behavior, especially during peak periods. Some surveys constitute a longitudinal group that was tracked over time. Others were a cross-section of commuters who were not tracked over time.

Other observations used survey data to calibrate choice models. These were then transformed into price elasticity and travelers VOT.

Opinion surveys were conducted to measure traveler satisfaction. An investigation of accident trends and their causes (i.e., weaving at entrances and exits) as well as for vehicle emissions was performed and compared to vehicle emissions along alternate roadways.

The report is divided into seven chapters. In chapter 2 the author examines the observed impact on traffic and travel behavior. Traffic counts, speeds, and observed vehicle occupancies are all documented for analysis. Chapter 3 further investigates the travel surveys to determine travel behavior. Trends in public opinions and survey results are addressed in chapter 4. Chapter 5 of the report deals with modes of traveler choice and decision-making about whether to use the toll lanes or not. This analysis also includes choices such as mode, time of day, and transponder possession. Corresponding price elasticity is also reported in this chapter.

In chapter 6 an analysis of collisions is performed and a comparison with another route (SR 57) is done to determine high accident rate locations. Chapter 7 of the report focuses on the emissions from vehicular travel on SR 91X and compares them to three alternative scenarios: (i) HOV lane substitution, (ii) GP lane substitution, and (iii) "no build" alternative."

The report is concluded in chapter 8 with the results and findings of the study performed by the author. The report concludes the analysis of the SR 91 express lanes with several key findings. One of the key results is that an increasing number of travelers are willing to pay a toll to use the Express Lane in SR91. However, even among those travelers who use the road, very few use it every trip and only do it as a necessity. Another key factor was that the demographic of the traveler who was most likely to use the Express Lane was female. High income, age, and commuting to work were all significant factors in the acquisition and use of transponders to use the 91 Express lane. Thus, women with high income use the toll lanes more often than men and other demographic groups, although all demographic groups have used the lanes at some point. The result of implementing the 91 Expresss lane has increased the capacity of the freeway.

The 91 Express now carries 1,400–1,600 vphpl, which is more than the capacity before the toll lanes were opened. Thus, the attraction of traveling on a relatively congestion-free road has induced more travelers to use the freeway compared to before the toll lane was operational. The 91 Express has received a largely favorable response from the general population of travelers; however, congestion has steadily increased in the region, especially with the opening of the ETR, where the eight lanes merge into six lanes when intersecting with 91 Express.

2.1.1.1 Katy Freeway in Houston Texas (nchrp_rpt_694)

The Katy managed lanes are HOT lanes that are located along Interstate 10 in Texas. The KATY HOT lanes are 12 miles long and are located in the median of the I_10 highway in Houston's Harris County.

The HOT managed lanes were launched previously as HOV lanes for vehicles carrying two or more passengers. These HOV lanes were then converted to HOT lanes in April 2009. The toll for vehicles traveling the entire length of the freeway is $4.00 each way during the peak hours of 7:00 a.m.–9:00 a.m. (eastbound) and 5:00 p.m.–7:00 p.m. westbound.

The KATY managed lanes are operated by the Harris County Toll Road Authority (HCTRA). HCTRA operates 100 miles of toll roads in the Houston area, including the Hardy toll road, which features fixed tolls collected manually and electronically.

The goals of the KATY freeway are listed in the NCHRP (report #694) as

(1) Not superseding toll rate covenants
(2) Maintaining investment grade rating of "A"
(3) Maintaining toll levels commensurate with toll rate policies associated with private toll road operators
(4) Allowing for maintenance and improvement of the HCTRA systematic

The HCTRA evaluates the performance of its managed lanes operation according to the guidelines developed by the USDOT in eight categories: (1) Traffic Performance, (2) Public Perception/Satisfaction, (3) Users, (4) System Operations, (5) Environment, (6) Transit, (7) Economics, and (8) Land Use. The Traffic Performance is measured using vehicle volume and mode share (SOV, HOV) metrics. The Public Perception is measured using acceptance of system as fair. The System Operations are measured by total transactions, toll revenue, average toll, O&M cost, violations/fines, and collisions accidents.

FIGURE 2.2   I-95 Express Entry Point

(Source: Photo courtesy of NCHRP)

HCTRA has listed the area of streamlining its message signs as one possible area of improvement.

2.2 I-95 EXPRESS LANES

The I-95 Express HOT lanes are variable-price lanes that run along I-95 in Miami, Florida. 95 Express allows motorists the option to pay to use a less congested lane in order to reduce their travel times from I-395 (South entry) in Miami to

Golden-Glades interchange in North Miami County. Additionally, BRT, registered carpools with three or more occupants (HOV3+), and registered energy-efficient (electric) vehicles are allowed to use the HOT lanes free of charge.

I-95 Express is being constructed in two phases. The first phase is located entirely in Miami-Dade County and is already fully operational.

Phase 2 of 95 Express will extend the HOT lanes northward into Broward County with additional entry/exit points allowing for long-distance commuters to reduce travel times between counties. Phase 2 is expected to be completed in June 2014.

One of the primary goals of the 95 Express is to maintain a minimum speed in the express lanes at 45 mph or above. This speed provides near free-flow conditions for peak hour travel in both the northbound and southbound directions.

Currently, there are three points of entry for the 95 Express northbound lanes and five points of entry for the southbound lanes. The FDOT will convert a total of 21 miles of HOV lanes between I-395 in Miami and I-595 in Ft. Lauderdale. The project is being completed with a grant from the United Partnership Agreement of the USDOT. The goals of the I-95 Express CP lanes are:

(1) Minimize travel time and maximize throughput
(2) Maintain a minimum speed of 45 mph on the express lane to allow for free-flow condition
(3) Reduce congestion
(4) Provide trip time reliability for commuters
(5) Increase BRT ridership and carpooling
(6) Meet increasing future travel demand

2.3 PERFORMANCE MONITORING

The Florida Department of Transportation has installed 31 electronic devices throughout the I-95 express lanes system; these devices are capable of monitoring speed and volume at that location and reporting to the ETC computers in near real-time mode. The system has the ability to continue to report speed data even when one or more detectors are non-working or disconnected. FDOT monitoring system will also have the capability to select some of the detectors that are considered more reliable (due to smaller variance in errors) than others and use the information provided only by those devices for analysis. Volume data is typically collected at toll gantries. Crash data is also monitored by the FDOT along I-95 Express from police in incident reports.

Using AVO and volumes, FDOT calculates average throughput rates for the I-95 Express variable-priced managed lanes. The Florida Turnpike Enterprise (FTE) collects data on tolls and revenue along the I-95 Express. FDOT tracks maximum tolls and monthly revenue trends in order to report this information on the operations of the 95 Express systems.

Figure 2.3 shows the proposed expansion of the 95 express lanes from Golden Glades Interchange north to I-595 in Broward County, Fl. The Phase 1

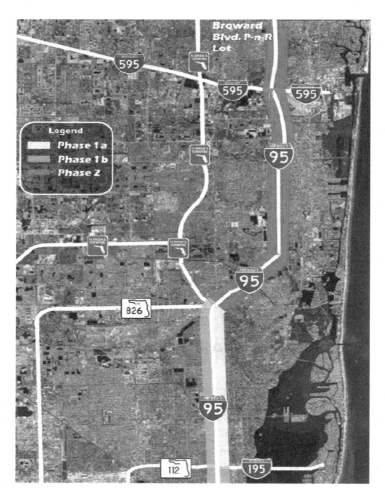

FIGURE 2.3   Express Lanes

(Source: Phase1 and Proposed Phase2. www.95express.com)

section is also shown (light gray). 95 Express is a multi-modal facility and the first in Florida. It handles over 290,000 vehicles per day volume, and this is expected to grow significantly over the next ten years.

95 Express was created from the I-95 freeway without the need for widening the roadway or adding any new lanes. Thus, infrastructure costs were kept to a minimum. The existing road was re-striped to add two new lanes, and the HOV lane was converted to HOT managed lanes. 95 Express has installed loop detectors and Intelligent Transportation System (ITS) infrastructure to monitor traffic speed and volume along the I-95. The 95 Express has also installed a transponder-based toll collection system (SUNPASS) for users to have tolls electronically deducted from their SUNPASS account. 95 Express operations have made a conscious decision not to try to maximize revenue as their primary goal. The primary stated goal of the 95 Express is to maximize throughput and relieve traffic congestion along the I-95 corridor in Miami.

95 Express installed operational tools to meet its stated objective of maximizing throughput with the ability to integrate with the tolling mechanisms, which change the tolls based on measured traffic density. The 95 Express Management Center FDOT District Six (see FDOT District Six 2009 Midyear UPA report) has developed and acquired software to meet the objectives and implement dynamic pricing capabilities. The dedicated software titled "Express Lane Watcher (ELW)" reads input from the 95 Express monitoring system to determine traffic density and demand throughout the system.

This real-time data is compared to historical data to calculate the tolls based on demand for the express lanes. The Express Lane Watcher (ELW) software uses an algorithm that enables it to follow the project objectives and generate toll rates, and these rates are updated every 15 minutes. Along with the ELW software, an operator is assigned to monitor the toll recommendations from the software and validate the recommendations throughout road conditions as determined by CCTV cameras, which are mounted along the corridor.

The ELW software has the ability to extract data for analysis and report on the performance of the 95 express lanes. The software also performs monitoring functions on the actions of the operator as the tolls are being updated.

The data collection effort is a combined effort between FDOT (district six) and FTE. Through the use of 29 detectors, the system is able to capture speed information along 95 Express. Volume data is recorded at the toll gantry when vehicles use their SUNPASS transponder to enter the 95 express lanes.

The FDOT District Six 95 Express Phase 1A Midyear evaluation report contains data presented as bar graphs for Speed Express Lane (EL) as well as GP lanes, travel times, and reliability.

Volume data, throughput, safety, and maximum revenue are also reported for each midyear report starting in 2009.

This information is widely available on various websites, including www.95express.com and also on FDOT websites that link to District Six information.

As mentioned previously, tolls are calculated using the algorithm on the ELW software and shared through the FTE to the SUNPASS system so that SUNPASS knows how much to deduct for each non-exempt vehicle passing through the system. As mentioned earlier, exempt vehicles such as registered carpools and BRT are not charged a toll. FTE summarizes all the trip and toll information into reports, which are then presented to District Six and disseminated to the public through their website.

The 95 Express implementation plans provided a good source of lessons to use in the implementation of other express lanes around the country. The 2009 mid-year report (see FDOT District Six Phase1A 2009 Mid-year reports) outlines some important facts in terms of lessons learned in the rollout of the 95 Express. Some of these expressed in the report include:

Project: Defines a strong project vision. Having a clear understanding of the project's purpose and goals provides for consistent decision-making throughout. The regional long-distance commuter is identified as the target market for the 95 express lanes Establish Project Schedule: An aggressive schedule was adopted by the UPA. As a result, functions of planning, design criteria, and operations were completed simultaneously rather than in a singular manner.

Institutional Approach: Develop concept of operations early on. This concept-driven approach provides guidance for planning, design, and implementation of the express lane.

Project Management: The 95 Express involved professionals from a variety of agencies. It is important that in order for the project to progress smoothly, individuals are able to take directions from the project manager directly independent of which agency they were affiliated with.

Information Sharing/Technical Data: Technical issues involved in integrating transit in the program included facility access, circulation, bus operations, and procurement of new transit vehicles.

Tolls along the 95 Express HOT lanes vary between $0.25 and $7.25 at peak hour. The toll is set depending on the traffic density along the I-95 corridor at any point in time. The traffic density is continuously monitored with loop detectors, which provide near real-time data on traffic volume, speeds, and density. This information is then fed into a computer system, which calculates the toll based on the reported traffic density and change in density. Before a toll is charged, a final toll determination is made by a toll operator who has access to closed-circuit TV visual verification of road conditions.

2.4 EVALUATION AND PERFORMANCE MEASUREMENT

According to the National Cooperative Highway Research Program (see www.nchrp.gov NCHRP report#694), congestion pricing performance can be evaluated with several different measures.

Some of the main measures include Traffic Performance, Public Perception, Facility users, System Operations, Effect on Environment, Transit, Economics, and Land Use.

2.4.1 Traffic Performance

Traffic performances are probably the most obvious of the performance measures since these are directly related to the motorist's perception of whether there has been an improvement in congestion and travel time along the express lanes.

Traffic performance includes such measures as LOS, average speed, travel times, time saved (by using the express lanes as compared to the general-purpose lanes), VOT, AVO, vehicle volumes, and average delay.

2.4.2 Public Perception

Public Perception relates to the awareness level of the public to the general operation of the toll facility, the typical toll levels, and also the level of satisfaction that the pricing scheme is fair and effective in improving congestion along the corridor.

2.4.3 Facility Users

Facility Users includes such information as demographic of typical user, mode of transport (HOV, SOV, etc.), vehicle make, and frequency of facility use.

2.4.4 System Operations

System Operations include general measures such as toll revenue, O&M cost, average toll, number of transactions, safety, number of accidents, and percent of time facility available.

2.4.5 Environment

Environmental characteristics include (1) air quality and (2) noise.

2.4.6 Transit

Transit includes such factors as ridership, percent on-time arrivals, service quality, and reliability.

2.4.7 Economics

Several economic factors are used to evaluate performance of the congestion pricing scheme. Some of these are cost/benefit analysis, gross product for the region affected by the express lanes, commercial costs and prices, retail traffic, and sales.

2.4.8 Land Usage

Land usage factors include housing buyer's decision-making process as well as commercial facility location planning.

Information about the performance of the I-95 Express is made public through a number of websites. One of the major sources of information is www.95express.com, which provides monthly reporting on ITS monthly usage reports, 95 Express Performance Reports, as well as DMS signage reports.

Some factors of interest that are measured and graphed in the 95 Express Performance Evaluation reports include average speed and peak period speed (peak period defined as 6 a.m.–9 a.m. southbound and 4 p.m.–7 p.m. northbound). Weekend Speed and Express Lane Speed are also monitored as described and reported on a monthly basis.

Volume data is collected and presented for both the express lane and the GP lanes, Average weekday and weekend traffic volumes for the express lanes as peak period and maximum weekday traffic values are presented as bar graphs in the report. Toll Revenue values including monthly revenue and number of transactions are available for the public to review.

General Information available on the 95 Express websites indicate that

1. The express lanes serviced 1.7 M total vehicle trips in May 2012, bringing the total to approximately 54 million trips
2. The total toll revenue of $42.4 million for the year 2012

This review of the current literature has shown that various analytical approaches have been taken to solve the congestion pricing problem. The I-95 Express Phase I has been examined by researchers at the University of Florida using Genetic Algorithm (GA). To the best of our knowledge, there is no study of the I-95 Express Phase 2 congestion pricing, nor is there any study that has applied the Simulated Annealing SPSA or FDSA algorithms to the I-95 Express. These methods have been shown to produce better results for other applications, and thus it would be expected that these algorithms will again be a better approach for the 95 Express and produce superior results than the GA. This proposal intends to pursue the application of these algorithms and perform a simulation of their performance.

Congestion Pricing Models

3

When we decide to implement a congestion pricing model, we are forcing the consumer to make a choice between two (or more) alternatives: saving time by using the toll lanes with less traffic or saving money by using the free lane. Thus, the roadway is divided into two route alternatives or possibly two-time alternatives (if the entire road is tolled at peak hour and not tolled or tolled less at other times during the day). The consumer must make a discrete choice to determine which transportation option is best for them. Many models have been developed to analyze this discrete choice problem. A good review of discrete choice models is provided in Ben-Akiva and Lerman (1985).

3.1 DISCRETE CHOICE ANALYSIS

Route choice is often closely associated with mode choice since drivers that use the free lanes in a congestion pricing location may decide to change their mode to Express bus or carpool ride to take advantage of the HOT lanes. Again, the mode choice problem has been addressed in detail in the literature. Some good sources of reference on mode choice include Ben-Akiva and Richards (1975), Atherton and Ben-Akiva (1975), and Daly and Zachary (1979).

Since we know that the discrete choice model has econometric implications, it is useful to obtain a background of econometric discrete choice models. For a good treatment of econometric discrete choice models. In some problems, the choice to be made by the consumer includes a combination of two or more variables. Thus, for example, instead of choosing one of two routes, the consumer must choose from alternatives that are described by both time *and* route. Thus, each choice includes both the time of day and choice

 DOI: 10.1201/9781003358701-3

of toll or no-toll routes. These types of problems are formulated using the multidimensional choice model. A multidimensional set of alternative choices can be demonstrated by the following example. (See Ben-Akiva and Lerman, 1985.)

Suppose we have

$$R = \{r1, r2, r3, \ldots\ldots r(n)\} = \text{all possible route choices} \tag{2}$$

and

$$T = \{t1, t2, t3, \ldots\ldots.t(n)\} = \text{all possible modes of travel} \tag{3}$$

Then the set

$$RxT = \{(r1, t1), (r1, t2)\ldots\ldots(r1, tn)..(r2, t1), (r2, t2), \ldots\ldots\ldots(r(n), t(n)\} \tag{4}$$

represents the set of all possible of route/time combinations.

Multidimensionality extends to any number of dimensions. It could include (along with route choice and time) other variables such as mode (SOV Express Bus), final destination, and other travel attributes.

When the set of alternatives is multidimensional, the *Nested logit model* can then be applied to formulate and solve the problem.

3.2 UTILITY THEORY

The consumer, when faced with the choice of deciding on which route or alternative to take, must first assess the value of each selection to him or her and then after weighing the relative gains will choose the alternative that is worth the most to them. Thus, consciously, or subconsciously, the consumer places a value on choosing the toll lane versus a non-toll lane, and when that value is high enough, they will decide to select that alternative and reject the other alternatives. This concept of valuing alternative choices is considered utility.

The utility theory is closely related and may be considered a component of discrete choice analysis.

The utility theory defines a utility function ($U(x)$) that is dependent on a number of explanatory (independent) variables to assign a quantitative value to each particular selection. So, for example,

$$U(1) = c(1) + \alpha t(1) \text{ (general form of utility function)} \tag{5}$$

or

$$U(1) = -0.75 + 3t(1)............ > \text{utility of choice } 1 (\text{toll lane}) \qquad (6)$$

$$U(2) = 0.25 - 5t(2)............ > \text{utility of choice } 2 (\text{free lane}) \qquad (7)$$

where $t(1)$ = time of travel for route 1 and $t(2)$ = time of travel for route 2.

If $U(1) > U(2)$, then we say that the consumer (driver) is more likely to choose route 1 than route 2.

Utility theory has been well developed in the consumer choice literature, and several references are available to provide a good treatment.

Small and Verhoef (2007) present a formulation of the maximum utility as constrained optimization problem:

Max $u(x, X)$

subject to $y = px + P'X$

where
    x = choice under consideration
    X = Vector of all possible alternatives
    Y = income
    P = price of choice x
    P = price Vector

A review of the theory of expected utility theory is provided by Fishburn (2010). Utility maximization is addressed by Aleskerov et al. (2007). Many authors in the literature show the direct connection between utility theory and game theory in econometric modeling. Thus, one may be well served to apply game-theoretic methods to solving the transportation congestion pricing and route choice problems as well.

3.3 THE CELL TRANSMISSION MODEL

The cell transmission model (CTM) as developed by Daganzo (1993) is an analytic technique that models the highway as being divided into equal length sections called "cells." This allows for the modeling of traffic flow along the highway link (e.g., HOT lane) to be analyzed more precisely by writing equations for the flow into each cell and out of each cell at any point in time, t. The cells are numbered from i = 1 to N, starting from the upstream section, where the traffic enters the network link to the downstream ending cell, where the traffic leaves the link.

The length of the cell is chosen such that it is equal to the distance traveled by a typical vehicle (traveling at average speed) during one clock tick. The cell transmission model is robust in that the cells can vary from small lengths to distances in excess over 1 km (see Daganzo, 1994).

Thus, if we assume that the dynamic pricing is updated every 15 minutes (as on 95 Express), then the cell length will be approximately 12 miles.

The equation of state for each cell is

$$n_I(t+1) = n_i(t) + y_i(t) - y_I(t+1) \tag{8}$$

The above recursive equation states that the cell occupancy at time $t+1$ $(n_I(t+1))$ is equal to the occupancy at time t, $(n_i(t))$ plus the inflow into cell I during time $t,(y_i(t))$ minus the outflow from the same cell during time $t,(y_I(t+1))$. The cell transmission model assumes an "output cell" at the end of the link of infinite capacity and also a "source cell" at the input of the link with infinite capacity.

The cell transmission model defines three variables related to flow from one cell (i − 1) into the adjacent cell (i):

$n_{i-1}(t)$ the number of vehicles in cell i − 1 at time t,

$Q_i(t)$ the capacity flow into i for the time interval t,

$N_i(t) - n_i(t)$ the amount of empty space in cell I at time t.

Here, $N_i(t)$ = maximum number of vehicles that can be present in cell i at time t The flow $x_i(t)$ is then = $\min\{n_{i-1}(t), Q_i(t), N_i(t)-n_i(t)\}$
That is for each pair of adjacent cells:

$$X_i(t) \le n_{i-1}(t)$$
$$X_i(t) \le Q_i(t)$$
$$X_i(t) \le N_i(t) - n_i(t) \tag{9}$$

These are cell transmission constraints for pairs of adjacent cells on each link.

3.3.1 Telecommuting Option

When the congestion pricing strategy is selected (HOT lane, Express Lane, etc.), the driver is faced with making a choice between using the tolled lanes and using the free lanes. But there is also a third option that the driver has: trip cancellation. The driver may decide that he or she is unable to pay the toll *and also that they are not willing to spend the enormous amount of time in the free lane* since their time could be used more efficiently.

For workers who are required to provide a service due to an employment contract, they do not have the option to simply cancel their trip to work, and thus they may choose instead to telecommute once the decision has been made to place a toll on their route to work. This is an option that is not included in many models in the literature. Many route-choice models only include an analysis of the different routes available (tolled and untolled) and provide a congestion pricing analysis on the basis of these alternative routes (see Mokhtarian and Salomon, 1994). Although there are many studies in the literature that examine workers' choice to telecommute (see Bagley, 1994; DeSanctis, 1984), there only a few studies that look at drivers choosing to telecommute given that a previously untolled route has now been tolled (see Bernardino, 1993).

Drivers can choose to telecommute for a variety of reasons and congestion, or tolling is only one. Thus, the problem formulation must consider this factor when determining the route-choice behavior (see Hartman et al., 1991). Also of importance is the level of productivity while telecommuting as compared with the level of productivity while at the work location. These and other requirements are important considerations when the driver considers the adoption of the telecommuting option if it is available.

3.4 SOLUTION METHODOLOGIES

Before a solution method can be applied, we must have a well-defined and well-formulated problem. As mentioned previously, the congestion pricing problem can be formulated into a variety of optimization frameworks. These may take the form of maximizing toll revenue, maximizing throughput, minimizing travel time, or maintaining a constant speed along the managed lanes. In these types of problems, the objective function must be clearly defined with the constraints that are imposed by the physical environment (such as number of lanes available for use, width, and geometry of these lanes). In addition, the decision-makers may set their own constraints such as budgetary restrictions or time restrictions. In the event the problem is linear, appropriate linear programming formulations and methodology can be used to solve the problem.

These linear programming problems take the general form

$$\text{Max } Y = c_1\left(x_1\right) + c_2 x_2 + c_n x_n$$

subject to

$$X \leq k \, ,$$

where

Y	=	profit function
x(i)	=	explanatory (independent) variables for Y
X	=	vector of independent variables
K	=	constant

These types of linear programming problems can be solved using well-known Simplex Algorithms (see Hillier and Lieberman, 2001).

For many congestion pricing problem formulations, the problem is nonlinear and will need alternate solutions for nonlinear problem types.

Some methods to solve nonlinear programs include interior point methods and algorithms (see Benson et al., 2000). Other methods for solving nonlinear problems include branch and bound method (see Avriel, 2003).

3.5 NONLINEAR OPTIMIZATION PROBLEMS

In many real-world applications the formulation of the optimization problem can become very unmanageable due to the large number of variables and the presence of a nonlinear objective function or nonlinear constraints. In these cases, an exact solution to the problem is unlikely due to the length of time and number of calculations involved. One example of this type of problem is the traveling salesman problem (TSP). In this problem, a salesman must visit a fixed number of cities in the shortest possible amount of time or distance. For 10 or 20 nodes in the network this problem can be formulated and solved exactly in a reasonable amount of computing time. However, as the number of nodes increases, the problem becomes non-polynomial bounded (NP), and other heuristic approaches must be employed to approximate the solution. This is the case with the congestion pricing optimization problem.

In this section we will discuss some methods to approach an approximate solution to this problem. The literature has examples of applications of such methodologies as simulated annealing (SA), GA algorithm, LOQO (an interior point method), KNITRO, and SNOPT (a quasi-Newton algorithm), among others.

3.5.1 Nonlinear Optimization Solution Methods

In this section we will review and describe some of the more popular solution algorithms for large-scale, nonlinear optimization problems.

LOQO (see Benson et al., 2002) is a mathematical programming formulation as follows:

Minimize $f(x)$

subject to $h_i(x) > 0.I = 1, 2, 3,$ m

the LOQO algorithm involves derivatives; therefore, the functions $f(x)$ and $h_i(x)$ are assumed to be twice continuously differentiable.

LOQO allows for bounds on variables, equality constraints, and ranges on inequalities. The initial step in the LOQO algorithm is to add slack variables to the inequality constraints:

Minimize $f(x)$

subject to $h(x) - w = 0, \quad w > 0,$

where $h(x)$ and w are vectors representing the values of $h_i(x)$ and w_i

Denoting the Lagrangian multiplier for the system by y, Newton's method is employed to iterate to triple values of (x, w, y).

LOQO then computes step directions Δx, Δy, Δw and proceeds to a new point:

$$x^{(k+1)} = x^{(k)} + \alpha^{(k)} \Delta x^{(k)}$$

$$w^{(k+1)} = w^{(k)} + \alpha^{(k)} \Delta w^{(k)}$$

$$y^{(k+1)} = y^{(k)} + \alpha^{(k)} \Delta y^{(k)}$$

where the value $\alpha^{(k)}$
is chosen such that $w^{(k+1)} > 0$, and $y^{(k+1)} > 0$.

KNITRO is an interior point algorithm that uses quadratic programming and trust regions to solve the sub-problems at each iteration. A trust region strategy is a method to handle both convex and non-convex spaces.

The Sequential Quadratic Program (QP) is used to handle non-linearity in the problem. After obtaining the step (Δx, Δw) from vertical and horizontal steps, NITRO checks a pre-defined "merit function" to determine if it provides sufficient reduction in the function to proceed in this direction.

SNOPT is a sequential quadratic programming (QP) algorithm that seeks a solution through solving a sequence of quadratic programs. The problem is formulated and solved in two phases. In the first phase, Phase 1 is the feasibility phase where the problem of minimum constraint violation is solved. Phase 2 is the optimality phase where the values of x and w (of the original problem formulation) from the solution to Phase 1 is used as the starting point for Phase 2.

SNOPT works well when n-m is relatively small (in the hundreds), where n is the number of variables and m is the number of constraints.

3.6 AMPL SOFTWARE

The algorithms discussed previously can be implemented and tested using the AMPL programming language. AMPL is described simply as a modeling language for mathematical programming. It has a comprehensive framework to model large-scale linear and non-linear optimization problems with variables that are either discrete or continuous. AMPL is specifically suited to handle problems involving maximizing or minimization of algebraic expressions subject to constraints expressed as inequalities.

Other mathematical programming languages include LINDO, LINGO, CPLEX, and MPL. Spreadsheet optimizers such as EXCEL provide optimization routines for solving relatively small-scale linear programming (LP) and nonlinear programming (NLP) problems; however, the AMPL interface is far richer in the functionality, in the ease of use and entering of information, and also in its capacity to handle problems with the number of variables (n) and number of constraints (m) in the thousands.

KNITRO for Mathematica is a solver for large-scale, nonlinear optimization problems. It can handle a variety of applications from different industries.

Some examples of typical large-scale problems that are suitable for KNITRO for Mathematica include

- Mixed Integer Programs (MIP)
- Mixed Integer Nonlinear Programs (MINLP)
- Least squares (linear and nonlinear)
- Linear Program (LP) and Quadratic Programming

(For more information on KNITRO for Mathematica, visit www.wolfram.com)

In one study comparing and testing the LOQO, KNITRO, and SNOPT algorithms, the authors have used AMPL to formulate and run test problems. The results of speed to convergence on optimality were compared where a solution was attained.

AMPL language was utilized for this test since AMPL is the modeling language that provides second derivatives to the solver. Test problems included in the study (Benson et al., 2000) were equality-constrained quadratic program (QP), inequality-constrained quadratic (QP), mixed-constrained QP, unconstrained QP, as well as the same four categories for nonlinear programs (NLP)

Benson et al. (2002) summarize the results of their tests with the following conclusions:

- For problems that have all equality constraints the problem should not be converted to inequality constraints and solved with algorithm such as LOQO. It should be solved directly.
- On the other hand, when the problem has all inequality constraints, the problem should be solved with an interior-point algorithm such as LOQO.
- When the constraints are mixed, no clear-cut guidance as to which algorithm to use exists.
- SNOPT works best on a reduced space of the variables feasible solutions, and thus it is efficient in solving problems with small degrees of freedom (i.e., (n-m) is relatively small).

The authors conclude the tests of these three algorithms with the observation "So far the results on the efficiency of state-of-the-art algorithms are quite encouraging." Other simulation-based algorithms exist in the literature that are significant to discuss for possible application to this problem.

Two of these are fairly similar and described by Kleinman et al. (1999). The first is called the Finite Difference Stochastic Approximation Algorithm (FDSA), and the second is called the Simultaneous Perturbation Stochastic Approximation (SPSA).

3.7 GENERAL DESCRIPTION OF STOCHASTIC APPROXIMATION

For a detailed treatment of stochastic approximation, see Kleinman et al. (1999). In this section we introduce FDSA algorithm and SPSA algorithm. Let

$$L(\theta) = E\left[f(\theta, \omega)\right] \tag{10}$$

be the function we want to minimize and ω be the representation of randomness in the system. F is a performance measure. And let $g(\theta)$ be the gradient of L with respect to (θ); stochastic approximation proceeds by finding a local optimum $(\theta*)$ by starting at $(\theta^\wedge)$ and performing iterations according to the following schedule:

$$\theta_{k-1}^\wedge = \theta_k^\wedge - a_k g_k\left(\theta_k^\wedge\right), \tag{11}$$

where g_k is an estimate of the gradient g, and a_k is a sequence of positive scalars such that

$$\sum a_k = \infty \tag{12}$$

The difference between the FDSA algorithm and the SPSA algorithm lies in the way the g_k is defined.

3.7.1 FDSA Algorithm

For FDSA algorithm the *lth* component of the gradient is defined as

$$\hat{g}_k\left(\theta_k^\wedge\right) = \left(y_{kl}^+\right) - \left(y_{kl}^-\right)/2c_k, \tag{13}$$

where c_k represents a sequence of positive scalars such that $c_k > 0$ and

$$\sum a_k^2 c_k^{-2} \leq \infty \tag{14}$$

and y_{kl}^+ and y_{kl}^- 1 represent components of noise in the loss function and are measured as

$$y_{kl}^{+/-} = f\left(\theta_k^\wedge\right) + / - c_k e_l \omega_{kl}^{+/-} \tag{15}$$

for $l = 1, 2, p$

Also e_l *represents the lth unit vector, and* $\omega l^{+/-}$ *represents randomness in the system.*

On the other hand, in the SPSA algorithm, which was developed by Spall (1987), the gradient function is defined in this manner:

let Δ_k element of R^p be a vector p of mutually independent random variables, such as for example Bernoulli (+/− 1) random variables each with probability of outcome of one half.

The lth component of the gradient is

$$g_{kl}^{\hat{}}\left(\theta_k^{\hat{}}\right) = \left(y_{kl}^+ - y_{kl}^-\right)\big/2c_k\Delta_k \tag{16}$$

In this case there are only two measurements of the loss function required to obtain one estimate of the gradient (as opposed to 2p measurements for the FDSA algorithm). The authors (Kleinman et al., 1999) then perform a test of performance of the FDSA and SPSA algorithms using Common Random Numbers (CRN). The authors describe CRN as a method that is used to reduce the variance of difference estimates in stochastic optimization problems.

The gradients of the objective functions are often determined by these differences as generated by the CRN technique.

The authors conclude that the method of using CRN will increase the rate of convergence for both the FDSA and SPSA algorithms to achieve optimality at a faster rate in significantly less time and computations.

The study also demonstrates that when the number of iterations of function evaluations for the FDSA and SPSA algorithms is equal, the SPSA algorithm gives smaller errors than the FDSA algorithm.

Kleinman et al. (1997) describe an application of the SPSA algorithm to the optimization of air transportation network by minimizing the gate delay for a multi-airport network.

The paper discusses the strategy of holding airplanes at the gate for a short period of time (gate delay) in order to avoid the more costly airborne congestion delay costs. This airline network consists of thousands of flights, and the task of determining how much gate delay is optimal can be very prohibitive.

The author demonstrates how the SPSA algorithm can be used to process gate delay costs from a simulation package (SIMMOD) and produce optimal gate holding strategies. Air traffic delay consists of three components: (i) gate delay, (ii) delay while taxiing, and (iii) airborne delay.

The problem of assigning the correct amount of gate delay to each flight in a network of thousands of daily flights is a large-scale, nonlinear optimization problem.

Software packages exist for simulating various portions of the operations of the flights in the network, including macroscopic, and mesoscopic detail, but these simulations do not provide an optimal solution independently. These simulation packages, such as SIMMOD, simply input a user-defined gate-holding policy and process the information given by the user about the network to output the corresponding delay costs for that policy. Other algorithms such as FDSA are also used to solve this type of large-scale optimization problem; however, SPSA has been shown by the authors to be superior to FDSA in that it only requires two measurements of the objective function in order to estimate the gradient, while FDSA requires 2p measurements to estimate the gradient, where p is the number of parameters.

In their paper Kleinman et al. (1997) set up the problem with four airport locations, with flights departing and arriving at all airport pairs except between airport# 2 and #4.

Each airport consisted of one gate and one runway. The structure of the network was input into SIMMOD for analysis.

The authors defined their objective function as follows:

$$L(\theta) = m_g(\theta) + 2.38 m_t(\theta) + 3.86\, m_a(\theta) \tag{17}$$

where $m_g(\theta)$ = total number of minutes of gate delay
$m_t(\theta)$ = total# of minutes of taxiing delay and
$m_a(\theta)$ = total number of minutes for airborne delay

The results of the report begin by assigning no gate holding $(\theta) = 0$ for each flight; then, after applying the SPSA algorithm, the total delay way improved in the network. The authors performed 20 runs with 30 iterations for each run.

The initial objective function value (total delay) was 8,796 while after applying the SPSA algorithm this total delay was reduced down to 7,618 (approximately a 13.4% reduction in total delay).

3.7.2 Simulated Annealing

Simulated annealing (SA) is a method for approximating the optimal solution for large-scale, nonlinear optimization problems.

SA takes its name from the physical process of annealing, where metals are treated and altered to achieve a certain desired shape.

In SA the solution is altered from one point to the next point until a near optimal solution is reached. The search for the desired solution proceeds by starting with an initial feasible solution and then moves to another point solution based on some search criteria. If this solution is better (as measured by the value of the objective function), then the search in this direction continues with another iteration. If the solution is worse, then a new gradient or search direction must be used, and a new solution point determined. In SA a probabilistic search method is used to determine the search path. This alteration process (annealing) continues until some stopping criteria are reached, at which point the "optimal" solution is said to be achieved. Convergence to the optimal solution is not guaranteed using SA.

3.7.3 Genetic Algorithms

Genetic algorithms (GAs) are a form of random search algorithms that do not use a derivative or gradient approach to arrive at the optimal solution. As mentioned previously, GAs work by simulating the biological processes of population evolution such as natural selection, mutation, crossover, and new solution (individual).

The process starts out with a set of possible solutions called "individuals" or parents. These parents are selected according to some fitness criteria (i.e., being a feasible solution to the optimization problem). The individual who is deemed the most fit is then "selected" to regenerate and produce "offspring" or new solutions. Random numbers are used to generate the pairs of "parents to produce new "offspring."

After a crossover process, and a subsequent mutation of the parent solution, a new solution or "offspring" is obtained. This process is then restarted, and the fitness of these new solutions is evaluated against the fitness criteria for providing new or better solutions.

This iterative process is continued until some pre-determined stopping criteria is achieved, and at this point the "optimal" solution is achieved. In many cases, a stopping criterion is defined simply by a fixed number of iterations of the algorithm, for example 100 iterations.

As mentioned earlier, with this type of random search algorithm, there is no guarantee of convergence to the global optimum, and any solution arrived at could be a local optimum.

3.8 TOLLING METHODOLOGY

The dynamic nature of demand on the HOT lanes requires careful consideration to develop the proper tolling methodology.

The main goal is to reduce congestion and travel times along the corridor for any given demand. Although we cannot directly change the total demand for the freeway, we can adjust and control the demand flow on a tolled lane using price elasticity.

3.8.1 Price Elasticity

The economic concept of price elasticity tells us how one variable will change in percentage as we increase or decrease another variable such as price.

Thus, price elasticity relates the demand for a commodity or service to its price.

It is well known that as price for the HOT lanes increases, this will discourage some users from traveling on this lane since it is not worth it for them to pay this cost for this service.

These users are mostly travelers who use the freeway for personal or non-work purposes. These users will instead choose to use the GP lanes, which are free at peak hours, and reduce the congestion in the HOT lanes for the premium users, who must be at a certain destination at a certain time.

Thus, in developing our tolling methodology we will use price elasticity to accurately control the flow of traffic on the HOT and GP lanes to near optimal values and minimize congestion.

Burris (2003) defines the price elasticity of travel demand as

$$E = \frac{(Q_2 - Q_1)/Q_1}{(P_2 - P_1)/P_1} \tag{18}$$

where
$\quad Q_2$ = demand level 2
$\quad Q_1$ = demand level 1
$\quad P_2$ = price level 2
$\quad P_1$ = price level

According to Burris (2003), the price elasticity of demand can be subdivided into several components including operating cost (fuel, oil, etc.), tolls, travel time, insurance costs, and parking availability costs.

The Victoria Transport Policy Institute (www.vtpi.org/elasticities.pdf), estimates that price elasticity due to tolls vary in the range from −0.1 to −0.4. This means that for every 10% increase in toll price, the demand will decrease in range between 1% and 4%.

For freeways such as the 95 Express, a price elasticity value of −0.25 is a reasonable estimate. The process of developing an optimum tolling scheme begins with first determining what the optimal flows are on the HOT and GP lanes; these flows can be determined for a given total demand D, for the highway such as to minimize travel time, T. As a result of determining the optimal split of traffic, one can then determine the base toll for the HOT lane for this demand using the toll equation developed by Hobeika (2003).

However, the actual traffic flows on the HOT and GP lanes may vary widely from this optimum, and thus we will assess an adjustment (either increase or decrease) in toll pricing to cause a desired change in the flow due to price elasticity.

The dynamic toll thus consists of two parts:

$$Toll = base(b) + percent(p)$$

Base toll can be determined from the travel time versus toll relationship developed by Hobeika.

Hobeika's toll formula is

$$Toll = VOT * (T_{congested} - TT) \tag{19}$$

where

$T_{congested}$ = Congested travel time as determined from travel time equation
TT = Free –flow travel time
VOT = Value of Time

The methodology is outlined below.

3.8.2 Toll Algorithm

The following steps are implemented to determine the optimal congestion price due to dynamic time-varying demand on I-95 Express HOT lanes.

1) Input total demand flow D (veh/hr) for each 15-minute period. Initialize toll = $0.50. Please note that there is a minimum toll on I-95 Express even if there is no traffic present.
2) Determine formula for travel time along HOT lane and GP lanes (use Akcelik equation). Denote total travel time along HOT lane as T_1 and total travel time along GP lanes as T_2.

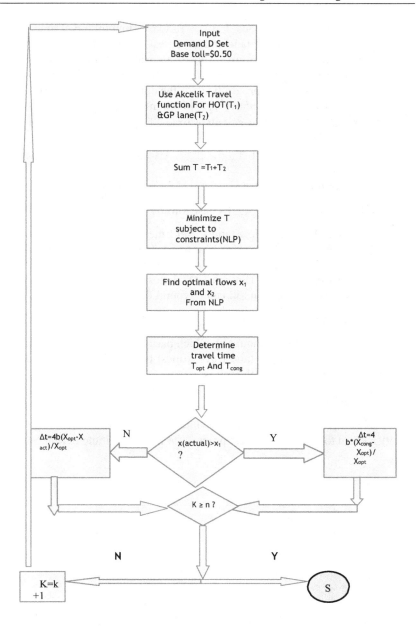

FIGURE 3.1   Process

3) Sum travel time functions $T = T_1 + T_2$ to obtain total travel time for all demand (both HOT lane and GP lanes) for period.

4) Use nonlinear programming (NLP) algorithm to minimize total travel time T subject to relevant constraints.

5) Solution of nonlinear program yields the optimal flows for the HOT lane (x_1) and the GP lanes (x_2), which minimize total travel time.

6) If x(actual) > x_1, then calculate time, $T_{congested}$ using travel time formula. Determine difference $T_{congested} - T(opt)$.

7) If x(actual) < x_1, and $T_{(opt)} > T_{actual}$, Use ($T_{(opt)} - T_{actual}$) to determine change in toll.

8) From formula for toll (see Lee and Hobeika, 2003),

$$toll = VOT * \left(T_{opt} - T_{free\text{-}flow} \right) = base\ toll\ (b)$$

Note that the actual flow (and travel times) on the HOT lanes can be greater or less than the optimal (since calculated optimal T is for the SUM of HOT lanes and GP lanes combined).

Thus if the actual flow on HOT lanes is greater than the optimal, then the change in toll can be found from

Change in toll $\Delta t = p = 4b * (T_{congested} - T(opt) / T(opt)$

This represents the amount by which base toll is to be increased.

If the actual flow (and travel times) on the HOT lanes is less than the optimal, then the toll is

change in toll $\Delta t = 4b * \left(T_{(opt)} - T_{actual} \right) / T(opt)$

This represents the amount by which the toll is to be decreased.

9) Return to step 1 and repeat process for next 15-minute period.

Model Formulation and Results

4

4.1 OBJECTIVE FUNCTION

The congestion pricing problem can be formulated in terms of either a maximization- or minimization-type problem. The actual form of the objective function is dependent on the goals of the authorities involved.

The goal is typically a maximization of throughput or minimization of travel times along the corridor. Maximization of revenue, although possible, is not a typical objective and in some cases may not lead to minimum congestion.

For our cases we will consider the objective to minimize travel time along the link.

4.1.1 Akcelik Travel Time Function

Several Freeway Travel-Time functions have been developed over the years. Two of these are

(1) Akcelik Travel Time function and
(2) BPR Travel Time function

The Akcelik Travel time function takes the form

$$t = t_0 + + \left\{ 0.25T \left[(x_{1k} - 1) + \left\{ (x_{1k} - 1)^2 + \left(\frac{8Jax_{1k}}{QT} \right) \right\}^{0.5} \right] \right\} \tag{20}$$

DOI: 10.1201/9781003358701-4

where

t	=	average travel time per unit distance (hours/mile)
t(0)	=	free-flow travel time per unit distance (hours/mile)
T	=	flow period, i.e., the time interval in hours, during which an average arrival (demand) flow rate, v, persists
Q	=	Capacity
x	=	the degree of saturation, i.e., v/Q
Ja	=	the delay parameter.

Figure 4.1 shows the Akcelik equation. Table 4.1 and Figure 4.2 shows the road classes, and they are

(1) Road class 1: Freeway, (2) Road class 2: Arterial 1, (3) Road Class 3: Arterial 2
(2) Road class 4: Rural 2-Lane Hwy, and (5) Road class 5: Rural 1-lane I-95 Express would be classified as a Class 1 road for this research

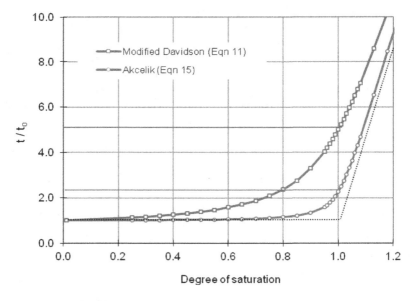

FIGURE 4.1   Akcelik Travel Time Function

(Source: www.sidrasolutions.com)

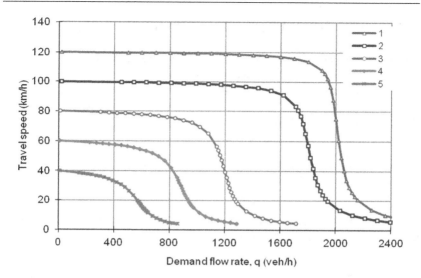

FIGURE 4.2   Speed versus Demand for Five Road Classes

(Source: sidrasolutions.com)

TABLE 4.1   Model Parameters of Five Road Classes

	ROAD CLASS	1	2	3	4	5	AKCELIK
Zero-flow speed given >>	v_0 km/h	120	100	80	60	40	80
Flow (analysis) period given >>	T_f h	1.00	1.00	1.00	1.00	1.00	1.00
Capacity (max. flow) given >>	Q veh/h	2000	1800	1200	900	600	800
Zero-flow travel time >>	t_o h/km	0.00833	0.01	0.0125	0.016667	0.025	0.0125

(*Source:* www.akcelik.au)

4.1.2 BPR Travel Time Function

The second travel time function that we consider is the Bureau of Public Roads (BPR) equation. This takes the following form.

The standard equation for the BPR curve is

$$\text{Congested Time} = \left[\text{Free Flow Travel Time}\right] * \left[1 + 0.15 * (v/c)^4\right] \qquad (2)$$

$$\text{Or} \quad \mathbf{T} = \mathbf{T_0}\left[1 + 0.15 * (v_k/c)^4\right],$$

where v/c = Volume/Capacity Ratio.

We choose to implement the Akcelik travel time function for this research proposal due to its applicability to the 95 Express.

4.2 PROBLEM FORMULATION

The goal of the 95 Express is to minimize the total travel time along the I-95 corridor for both the express lanes (EL) and the general-purpose lanes (GP).

To express this mathematically we will sum the travel times for ALL the demand flow entering the corridor:

$$\text{Thus Total travel time} = \sum_0^{x1} t_1(x1) + \sum_0^{x2} t_2(x2) \qquad (3)$$

We then substitute the travel time equations into the equation above to obtain the following travel time optimization problem, including cell transmission constraints.

Since we are formulating the general model, we will consider a multi-period time horizon with time periods numbered from k = 1 to n. Each time period represents a 15-minute interval on the HOT lanes.

$$\text{Minimize } T = \sum_{k=1}^{n}\left[\sum_0^{x1} t_0 + +\left\{0.25T\left[(x_{1k}-1)+\left\{(x_{1k}-1)^2+\left(\frac{8\text{Jax}_{1k}}{QT}\right)\right\}^{0.5}\right]\right\} + \sum_0^{x2} t_0 \right.$$

$$\left. +\left\{0.25T\left[(x_{2k}-1)+\left\{(x_{2k}-1)^2+\left(\frac{8\text{Jax}_{2k}}{QT}\right)\right\}^{0.5}\right]\right\}\right]$$

Subject to constraints:

$$x_{1k} + x_{2k} = D_k \qquad\qquad \textbf{\textit{for}}\ \forall\ \textbf{\textit{k}}$$

$$x_{1k} > 45\rho_{1k} \qquad\qquad \textbf{\textit{for}}\ \forall\ \textbf{\textit{k}}$$

$$x_{1i(k)} \le n_i - 1(k) \qquad\qquad \textbf{\textit{for}}\ \forall\ \textbf{\textit{k}}$$

$$x_{1i(k)} \le Q_{i(k)} \qquad\qquad \textbf{\textit{for}}\ \forall\ \textbf{\textit{k}}$$

$$x_{1i(k)} \le N_{i(k)} - n_{ik} \qquad\qquad \textbf{\textit{for}}\ \forall\ \textbf{\textit{k}}$$

$$n_i(k+1) \le n_i(k) + x_{ik} - x_i(k+1) \qquad\qquad \textbf{\textit{for}}\ \forall\ \textbf{\textit{k}}$$

$$x_{2i(k)} \le n_i - 1 \qquad\qquad \textbf{\textit{for}}\ \forall\ \textbf{\textit{k}}$$

$$x_{2i(k)} \le Q_{i(k)} \qquad\qquad \textbf{\textit{for}}\ \forall\ \textbf{\textit{k}}$$

$$x_{2i(k)} \le N_{i(k)} - n_{ik} \qquad\qquad \textbf{\textit{for}}\ \forall\ \textbf{\textit{k}}$$

$$x_{1k} \ge 0$$

$$x_{2k} \ge 0$$

$$N_{ik} \ge 0$$

$$n_{ik} \ge 0$$

$$Q_{ik} \ge 0$$

where x_{1k} and x_{2k} are the flows in time *period k* on the Express lane and GP lane respectively, and D is the total demand.

The variables x_{1ik} and x_{2ik} are flows in route 1 (express lane) and route 2 (GP lane) in *cell I* in time period *k* respectively. These are cell transmission variables.

N_{ik}, n_{ik}, and Q_{ik} are the maximum occupancy in cell *I* at time period k, the actual occupancy in *cell I* in time *period k*, and the maximum flow capacity into *cell I* in time period k respectively. Please note that in our application, we consider the I-95 HOT lane to consist of a single cell, with a source cell before and output cell following.

The first constraint is a statement of the conservation of flow. Thus, the total demand is equal to the sum of the flows along the express lanes and the

GP lanes, combined. The second constraint is due to the FDOT speed requirement, which dictates that speed on the express lanes must be at least 45 mph.

The entering demand is assumed to be known for each 15-minute period.

4.2.1 Basic Assumptions

In formulating the model the following basic assumptions were used to facilitate the analysis:

1. Demand Flow D, is known for each 15-minute period
2. Density is measured by loop detectors and updated every 15 minutes
3. No queuing occurs on the express lanes (we constrain the speed to be >45 mph)
4. The capacity does not change during the period (Q = 2000 vphpl)
5. The same number of vehicles exits any access point as enters it; thus, the flow along the corridor is the flow at entry point

The nonlinear function minimization is carried out using MATLAB. The results are the optimal flow values (x_{1k} and x_{2k}) for the express lane (EL) and GP lane, which minimize the total travel time along the corridor given the constraints.

Knowing the optimal flow values we calculate the travel time along the express lane.

Since travel time is related to toll amount, we can calculate the appropriate toll value using the following equation:

Travel time t = TT + toll / VOT

where TT = free – flow travel time
VOT = Value of time. This value is usually taken to be \$15–\$25/hr depending on location. In this study we use VOT = \$20/hr.

We perform the function minimization for demand values ranging from 2,000 vph to 12,000 vph and determine the travel times. The toll is then calculated for each combination of demand and density, which minimizes total travel time. If this toll amount is less than \$.25, we round up to the next \$.25 since toll increments are in values of \$.25.

Figure 4.3 shows average toll rates during the year 2010 as reported by www.95express.com for the months January to June on I-95 Express HOT lanes.

The grey lines on Figure 4.3 plot average toll for peak period in the corresponding month, and the white lines plot the maximum tolls during the same months.

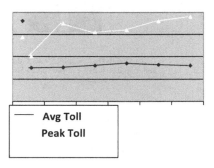

— Avg Toll
Peak Toll

Jan	Feb	Mar	April	May	June
$4.00	$3.00	$2.00	$1.00	$0.00	

FIGURE 4.3   Average Toll 95 Express

4.3 RESULTS AND SAMPLE CALCULATIONS

The following figures (Figures 4.4, 4.5, and 4.6) are the function minimization plots for a demand value of 3,000 vph.

4.3.1 Simulated Annealing Example

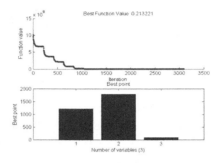

Demand = 3000vph. Cap(EL) = 4000,density = 25
toll = $0.04<$0.25 i.e. raise toll by $0.25
Toll = $0.50+$0.25=$0.75

Demand	Capacity (EL)	Optimal HOT Flow(x1)	Optimal GP flow(x2)	Density	Base Toll
3000	4000	1217	1782	25	$0.75

FIGURE 4.4   SA Example Calculation

4.3.2 Genetic Algorithm Example

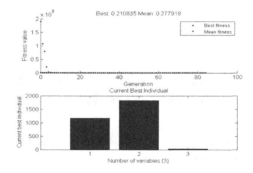

Demand	Capacity (EL)	Optimal HOT Flow(x1)	Optimal GP flow(x2)	Density	Base Toll
3000	4000	1175	1825	25	$0.75

FIGURE 4.5   GA Example Calculation

4.3.3 FDSA Example

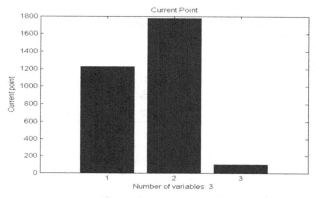

Demand	Capacity (EL)	Optimal HOT Flow(x1)	Optimal GP flow(x2)	Density	Base Toll
3000	4000	1225	1775	25	$0.75

FIGURE 4.6 FDSA Example Calculation

The optimal flows as generated by the three algorithms for the demand range of 3,000 vph to 14,000 vph are shown in Table 4.2.

TABLE 4.2 Optimal Flows by Algorithm, Demand

TOTAL DEMAND VEH/HR	CAPACITY (EL) VEH/HR	ALGORITHM	HOT FLOW (X1) (OPTIMAL)	GP FLOW (X2) (OPTIMAL)	BASE TOLL ($)
3000	4000	SA	1217	1782	0.75
3000	4000	GA	1175	1825	0.75
3000	4000	FDSA	1225	1775	0.75
6000	4000	SA	1988	4012	1.00
6000	4000	GA	1892	4108	1.00
6000	4000	FDSA	1800	4200	1.00
10000	4000	SA	3475	6525	1.50
10000	4000	GA	3460	6540	1.50
10000	4000	FDSA	3375	6625	1.50
12000	4000	SA	4550	7450	4.00

(Continued)

TABLE 4.2   (Continued)

TOTAL DEMAND VEH/HR	CAPACITY (EL) VEH/HR	ALGORITHM	HOT FLOW (X1) (OPTIMAL)	GP FLOW (X2) (OPTIMAL)	BASE TOLL ($)
12000	4000	GA	4525	7475	3.75
12000	4000	FDSA	4000	8000	3.75
14000	4000	SA	5770	8230	7.00
14000	4000	GA	5175	8825	7.00
14000	4000	FDSA	5175	8825	7.00

4.4 APPLICATION OF TOLLING ALGORITHM

The tolling algorithm previously developed can now be applied to the demand values of D = 3000, 6000, 10,000, 12,000, and 14,000 as shown in the above examples to determine the final toll for optimal minimized travel time.

Having applied the stochastic approximation algorithms to solve the problem, the next step in the toll algorithm is to determine the adjustment to the toll based on the deviation of the actual flow on the HOT lane from the optimal flow.

The calculations for the demand levels mentioned above and simulated actual flows are shown in Table 4.2.

In this table, each value for total demand is repeated once to signify two different dates or times when the same demand is observed with a different value for the actual flow on the HOT lane. This actual HOT flow value is used to calculate the deviation ($X_{congested} - X_{opt}$), which is then used by the toll algorithm to determine the adjusted toll.

For the first four rows in Table 4.3, the actual and optimal flow ratio x/Q (Q=capacity HOT lanes =4000 vph) is less than 0.6. Thus, there is no congestion and t = t (0). No need to adjust toll. *For the last row, flow is greater than capacity; do not reduce toll, as this will increase congestion.*

TABLE 4.3 Application of Toll Algorithm

TOTAL DEMAND D	OPT. HOT FLOW	ACTUAL HOT FLOW	BASE TOLL (B)	ADJUSTED TOLL P	FINAL TOLL T ($)
3000	1217	1520	0.75	n/a	0.75
3000	1217	1295	0.75	n/a	0.75
6000	1988	2150	1.00	n/a	1.00
6000	1988	1750	1.00	n/a	1.00
10000	3475	3600	1.75	0.2431	1.99
10000	3475	3525	1.75	0.09929	1.85
12000	4000	4500	4.00	1.7778	5.78
12000	4000	3750	4.00	−1.0667	2.94
14000	5770	5800	7.00	0.1448	7.15
14000	5770	5500	7.00	n/a	7.00

4.5 COMPARATIVE EVALUATION OF GA, FDSA, AND SA ALGORITHMS

The three algorithms used to solve the nonlinear program (NLP) were evaluated using statistical analysis methods to determine if there are any significant differences between the optimal flows recommended by each and the travel times produced by applying each solution.

The problem was constructed as a two-way ANOVA using travel time as the dependent variable and ALGORITHM and DEMAND LEVEL as the explanatory variables.

4.5.1 Vissim Simulation

The algorithm was tested in VISSIM MODELING SOFTWARE, using the simulation menu option. We manually input the demand and flow values from each algorithm to determine the simulated travel time output.

PTV-VISSIM is a transportation software tool used to model travel along a transportation network. Within the VISSIM network, one can model the flow along highways, arterials, intersections, and roundabouts. VISSIM also allows the modeling of the surrounding environments such as buildings, parks, and pedestrian traffic.

VISSIM allows the user to define the distribution of traffic, which is allowed on any link in the network. Thus, for example, one can define a portion of a network in such a way as to allow trucks from driving in certain lanes. See Figure 4.7, for example.

In VISSIM, one can simulate the flow of traffic on a highway by first defining a network using the option to construct a set of links. The links can be connected by using "connectors," which originate at one link and end at another link. Vehicle types are defined, and proportions of each type are also defined for each link as well as the vehicle speed distribution. Figure 4.8 shows a VISSIM vehicle input menu.

VISSIM has menu bar at the top of the user interface with several menu options.

One of these options is the simulation menu.

Within the simulation menu one can set up and run a simulation of a network for a specific period of time. Figure 4.9 shows the simulation menu.

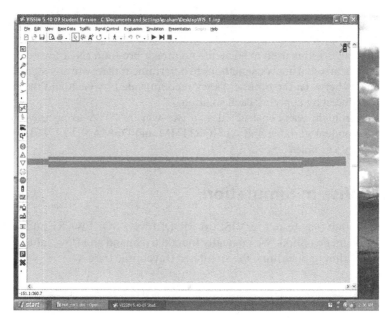

FIGURE 4.7   VISSIM Representation of HOT and GP Lanes

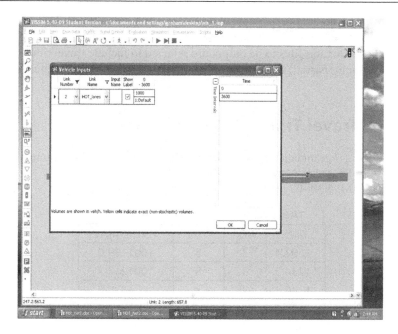

FIGURE 4.8 VISSIM Vehicle Inputs

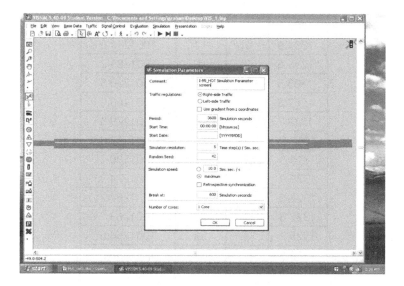

FIGURE 4.9 Vissim Simulation Menu

Once the simulation parameters are selected, the simulation can begin by selecting "Continuous" or "Single Step" run.

The evaluation files to be output are selected to include travel time.

A sample (partial) travel time output file from VISSIM is shown in Table 4.3(b) below.

4.5.2 Travel Times Data Collection

File: c:\programdata\ptv_vision\vissim530\examples\test1.inp Comment: Date: Thursday, May 23, 2013 9:11:27 PM
VISSIM: 5.30–09 [30156]

TABLE 4.3(B)   Output File from VISSIM

TIME	NO.	VEH	VEHTY TRAV
1202.4	3	5	1001189.0
1234.3	3	8	1001217.7
1246.3	3	3	1001236.5
1260.0	3	17	1001237.6
1266.0	3	43	2001222.5
1273.9	2	1	1001263.5
1281.2	3	44	1001237.3
1285.5	2	2	1001274.8
1293.9	2	12	1001273.6
1295.6	2	4	1001281.2
1305.5	3	94	1001230.5
1307.8	2	31	1001269.9
1313.7	2	28	1001279.5
1317.9	3	18	2001293.8
1319.5	3	25	1001290.2
1320.9	3	51	1001274.3
1322.3	3	67	1001264.2
1323.8	3	70	1001263.7
1325.2	3	88	1001254.0
1326.5	3	83	1001256.5
1327.8	3	26	1001295.0
1329.2	3	47	1001283.1
1330.7	3	96	1001255.4

TIME	NO.	VEH	VEHTY TRAV
1331.5	3	23	1001303.9
1332.1	3	126	1001230.1
1332.7	3	48	1001286.8
1336.5	3	10	1001318.5
1337.9	3	20	1001313.2
1338.1	2	32	1001299.4
1348.3	3	137	1001239.0
1352.4	3	190	1001188.5
1366.2	3	216	1001179.4
1368.4	3	16	1001346.2
1369.8	3	22	1001343.0
1371.1	3	41	1001328.8
1373.2	3	203	1001194.8
1379.0	2	7	1001360.7
1380.1	3	183	1001221.5
1381.3	3	205	1001200.9
1386.4	2	34	1001346.5
1397.5	3	221	1001206.4
1423.6	3	84	1001352.4
1425.4	2	6	1001409.1
1426.8	3	92	1001352.7
1426.8	2	9	1001408.7
1423.6	3	84	1001352.4
1425.4	2	6	1001409.1
1426.8	3	92	1001352.7
1426.8	2	9	1001408.7
1428.2	3	100	1001348.8
1428.2	2	13	1001406.3
1429.5	3	104	1001346.7
1430.9	3	125	1001328.8
1432.4	3	144	1001313.4
1437.4	2	11	1001416.1
1438.9	2	14	1001416.1

(Continued)

TABLE 4.3(B) (Continued)

TIME	NO.	VEH	VEHTY TRAV
1440.3	2	19	1001414.7
1441.7	2	27	1001406.6
1443.1	2	37	1001401.3
1444.4	3	59	1001390.2
1445.9	3	81	1001377.9
1447.4	2	29	1001410.5
1447.9	2	15	1001423.8
1448.7	2	39	1001405.5
1449.4	2	24	2001421.4
1449.5	3	36	1001328.8
1450.1	2	45	1001404.2
1450.8	3	163	1001312.2
1451.3	2	38	1001408.2
1451.6	2	60	1001396.6
1452.2	3	86	1001307.5
1452.7	2	40	1001408.2
1453.0	2	62	1001396.7
1453.4	3	255	1001223.5
1453.6	3	235	1001245.0
1453.7	3	139	1001266.4
1454.1	2	49	1001406.4
1454.4	2	42	1001409.2
1454.4	3	80	1001386.7
1455.1	3	254	1001226.3
1455.6	2	54	1001406.4
1455.8	3	120	1001357.5
1455.8	2	52	1001407.5
1455.9	3	288	1001198.1
1457.1	2	56	1001406.2
1457.2	2	55	1001406.3
1457.2	3	21	1001354.8
1458.5	2	46	1001411.9

As an example, the simulated annealing algorithm was initially used to solve the non linear programming problem. Thus, we use the optimal result from this algorithm in the first simulation to find the simulated travel times and costs for the I-95 express (HOT) lanes and the GP lanes.

The test plan is shown in Table 4.4, including sample data:

The same process is performed for GA, and the results are recorded in Table 4.5.

Finally, the values and results for the simulated annealing algorithm are also recorded in Table 4.6.

The results from the three algorithms are then compared to determine the best choice for implementation and also for use in the statistical analysis.

TABLE 4.4   Travel Time Data Collection Table: FDSA

DEMAND D (VPH)	HOT FLOW (X1)	GP FLOW (X2)	TRAVEL TIME
2000	1000	1000	25445.45
3000	1225	1775	38838.11
4000	1690	2310	53980.51
5000	1750	3250	65987.44
6000	1850	4150	83461.9
13000	4750	8250	275414.78
14000	5175	8825	290450.04
15000	5625	9375	333229.16
16000	5850	10150	342491.29

TABLE 4.5   Travel Time Data Collection Table GA

DEMAND D (VPH)	HOT FLOW (X1)	GP FLOW (X2)	TRAVEL TIME
2000	1000	1000	26848.31
3000	1175	1825	38482.09
4000	1690	2310	53117.18
5000	1750	3250	65628.14
6000	1892	4108	83461.9
13000	4750	8250	290596.35
14000	5175	8825	304179.19
15000	5625	9375	333229.16
16000	5850	10150	342491.29

TABLE 4.6 Travel Time Data Collection Table: Simulated Annealing

DEMAND D (VPH)	HOT FLOW (X1)	GP FLOW (X2)	TRAVEL TIME
2000	800	1200	24931.39
3000	1217	1782	38479.42
4000	1498	2502	53103.55
5000	1580	3420	67095.93
6000	1850	4150	83461.9
13000	4818	8182	312855.92
14000	5770	8230	294573.57
15000	5346	9654	319565.91
16000	5975	10025	343296.05

4.6 ANALYSIS OF VARIANCE

The results of the three algorithms were analyzed using ANOVA to determine if there is any difference in the travel time or cost. We want to determine the minimum travel time and the method that produces this result to a significant (95%) level.

ANOVA is a statistical technique that can be used to find the analysis of variance between and within groups of three or more.

One-way ANOVA can be used with three groups (three algorithms) to test the travel times.

The ANOVA method is related to the estimation of how much variance there is in the population. The actual variance in each population may be unknown, but this can be estimated by sampling the population and using the sample variance as the estimate of the population variance.

ANOVA compares the differences in the sample variances to find out if there are any statistically significant differences between the group samples.

4.6.1 Two-Way ANOVA

The two-way ANOVA results are shown below (see Appendix A for ANOVA worksheet data): *Table 10.2-way ANOVA*

TABLE 4.7   Two way ANOVA results for Travel Time

SOURCE	DF	SS	MS	F	P
ALGORITHM	2	4.87951E+12	2.43975E+12	3.49	0.034
DEMAND	8	1.21419E+16	1.51774E+15	2171.30	0.000
INTERACTION	16	2.98281E+13	1.86426E+12	2.67	0.001
ERROR	108	7.54919E+13	6.98999E+11		
TOTAL	134	1.22521E+16			

Individual Value Plot of TRAVEL TIME versus ALGORITHM, DEMAND Welcome to Minitab, press F1 for help.

Retrieving project from file: 'C:\Documents and Settings\Computer\My Documents\hot_95.MPJ'

Figure 4.10 shows that for demand levels between 2000 and 6000 vph, the three algorithms behave quite similarly. However, for higher demand levels (13,000 vph and above), the SA algorithm produces different results from the other two.

The results of the two-way ANOVA shows that the algorithm is significant in determining travel time. Also, the interaction between algorithm and demand is a significant factor. In general, one can conclude that when

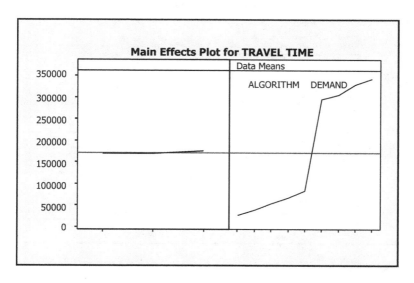

FIGURE 4.10   Main Effects Plot

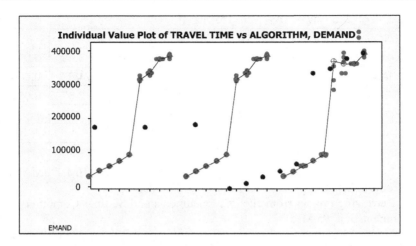

FIGURE 4.11 Travel Time versus Algorithm, Demand

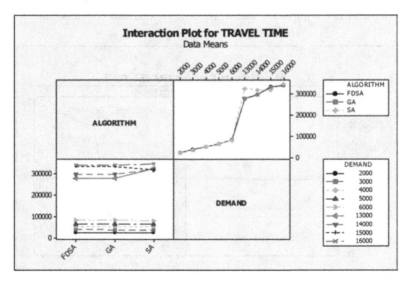

FIGURE 4.12 Interaction Plot

demand is greater than capacity, the FDSA and GA algorithms produce travel time results that are lower than the SA algorithm in solving this problem. Prior analysis showed that for demand values less than or equal to capacity, there is no significant difference between the algorithms in terms of their performance.

Data Collection and Demand Forecasting

5

5.1 DATA COLLECTION

The demand for usage of the 95 express lanes can be determined based on forecasted information. With the upcoming implementation and opening of Phase 2 of the 95 Express, it may be suitable to perform an analytic forecast to assess the demand for tolled express lanes in Broward County. Most forecasts currently in use have daily usage and demand data, which are used to forecast demand for future time periods.

The data collection efforts in this study consist of accessing data from three primary data warehouses for 95 Express. The FDOT District Six maintains data from its 31 detectors along the I-95 Express managed lanes on speeds and traffic densities. As mentioned earlier, the FTE collects information on the traffic volume based on tolled usage SUNPASS and exempt vehicles (such as hybrid-electric vehicles) at entry points to the 95 Express. A Third source of data is the STEWARD data warehouse maintained by the University of Florida on traffic for all districts within the state of Florida. This database can be queried, and reports can be obtained on different traffic patterns at different points along I-95 and other highways in Florida. A regular schedule of observational data of traffic patterns will also be recorded on travel times at different times (peak hour and off-peak hours) on the managed toll lanes at locations on the 95 Express.

Historical data of three years is used for this dynamic pricing study to calibrate the parameters of the algorithms and to assist with the dynamic

DOI: 10.1201/9781003358701-5

forecasting model utilized. With a dynamic pricing environment, where prices are changing every 15 minutes a daily forecast is not sufficient to provide accurate information for travel demand at different periods throughout the day.

The existence of loop detectors, which provide near real-time information, increases the flow of information and capabilities of operators to know what the typical demand scenarios are on an hourly basis.

Thus to capture the richness of this data and properly utilize this information in a forecast, a new forecasting technique that is capable of capitalizing on the wealth of data from the detection devices must be employed.

One such method, which has been utilized successfully in the econometric world for more than 20 years, is the Generalized Auto-Regressive Conditional Heteroskedasticity (GARCH) algorithm.

GARCH is a type of regression modeling that has been used in stock pricing and economic modeling. Developed by Robert Engle and Bollerslev, GARCH essentially removes restrictions on classical linear regression models. In particular, linear regression models assume that the error term has variance which does not change with time, i.e., stationary. However, in the real world, this is hardly ever true, and thus the classic linear regression model is always plagued by large errors due to the inaccuracies in the model.

GARCH assumes non-stationary error terms, hence its heteroskedasticity, and utilizes the large volume of real-time data to calculate extremely accurate forecasts for demand management. The 95 Express presents a good test-bed for the application of a forecasting algorithm such as GARCH and will benefit significantly from its improvement in accuracy in light of the proposed extension of the tolled dynamic pricing lanes into Broward County, Florida.

The expected demand for such services as 95 Express in this new region should be carefully assessed in order to determine the proper tolls and change in pricing for this new effort. It cannot be assumed that what happened successfully in Phase 1 of this project will necessarily carry over to similar success for Phase 2 of the project. An in-depth analysis should be performed with cost/benefit ratios to determine how best to proceed with the construction, implementation, and operation of Phase 2 of 95 Express.

GARCH has been applied successfully to applications outside of econometric models. It has been successful in improving forecast accuracy in supply chain systems, by reducing bullwhip effect and modeling demand with high level of accuracy as demand moved through the different stages of a multi-echelon supply chain system.

In one case, GARCH was applied to model a supply chain system demand for spare parts and was instrumental in reducing the forecast error from 21% down to 6.7%. These types of improvements can be transferred and successfully applied to travel demand for tolled express lane service along the I-95 in Broward County.

GARCH works well with the auto-ID technology, which is typically used along tolled roadways to collect the toll. In prior implementation of the GARCH algorithm, the data was read directly from Radio Frequency ID (RFID) tags in a point-of-sale environment and fed into the manufacturing warehouse setting to a dedicated computer system which was able to adjust the forecast dynamically in near real time. This type of dynamic updating is necessary for this dynamic pricing algorithm to determine optimal prices on the 95 Express toll road.

For a good description of an application of GARCH, see Datta et al. (2009). This paper addresses forecasting and risk analysis in supply chain management and discusses how reduced variation in forecasting errors can reduce the risk and costs associated with inaccurate forecasting information.

In terms of application of GARCH to 95 Express, one does not want to over-price the toll road based on inaccurate information from the data, since this will drive potential travelers away from using the express lane and reduce revenues.

Thus, tolls set too high or too low can result in reduced revenues and inefficient operations on the 95 Express toll road. It is important that the toll be set at the right level for the user who needs to get to their destination in a reasonable amount of time. The accuracy of the demand information is critical in setting the toll at the correct level and maximizing throughput and revenue in order to operate the express lanes profitably.

5.2 DEMAND FORECASTING

GARCH is a type of Vector Auto Regression (VAR) technique that is ideally suited for problems where there is a high level of volatility in the data. Examples include stock pricing where intra-day demand can vary widely, and corresponding prices can also change greatly on an hourly basis. GARCH is able to manage the complexity in such volatile environments where classic linear regression models would produce huge errors due to its inherent assumptions of stationarity. Vector Auto Regression is a multivariate model that allows for regression on multiple variables simultaneously.

Vector Auto Regressive models allow variables to be dependent not only on its own past values but also the past values of all values in the model.

The inaccuracies in classic linear regression models are derived from multiple sources, not only from unrealistic assumptions, but also from batch data (usually on a weekly or monthly basis), which are of poor quality and do not capture the large increases or decreases within the smaller time periods (hour by hour).

The standard GARCH (p,q) model with Gaussian shocks has the form

$$y_i = b_0 + bx' + \varepsilon$$

where $\varepsilon_t / \Psi_s = Normal\ 0, h_t$

and $h_t = \alpha_0 + \sum^q \alpha_i \varepsilon^2 - i + \sum^p \beta \varepsilon_i h_{(t-1)}$

Where $h_t = $ variance of the error ε_t

One method of estimating GARCH model parameters is by finding values that maximize the log-likelihood function:

Log - Likelihood function :

Likelihood function :

$$LF = 1/2 \sum (log(b_i) + \varepsilon^2 / b)$$

This GARCH process is described by $q + 1$ coefficients (α_i), p coefficients (β_i), as well as the endogenous/exogenous variables y_t and x_t.

Other types of GARCH models include asymmetric GARCH (AGARCH) and exponential (EGARCH).

GARCH model parameters can be estimated by maximizing the conditional log-likelihood function (MLE).

An initial approximation for the parameter vector is used to start the process, and numerical optimization is then applied to iterate to an acceptable solution.

The performance or accuracy of the GARCH forecast can be measured by the Statistics of Fit. Statistics of Fit examines how well the forecast performs by comparing the forecast with the actual data.

Some measures of performance include

(1) Mean Square Error (MSE)
(2) Mean Average Percentage Error (MAPE)
(3) Aikiki Information

Using data from the STEWARD database, we demonstrate how the GARCH model can be used to predict volatility in the transportation demand on I-95 Express.

The STEWARD database allows for the selection of a range of dates at a particular station along 95 Express to view or download the actual traffic volume as recorded by detectors. This data was downloaded for October 2012 and analyzed in EXCEL.

The following graph shows the daily volume for October 27, 2012 at the station located at the I-95 NB HOT ramp at NW 46th St. (see Appendix for data).

The Weighted Moving Average (WMA) forecast was developed from this daily demand to predict demand profile to predict demand for future day. The hourly demand is sub-divided into four 15-minute periods, and the period # is

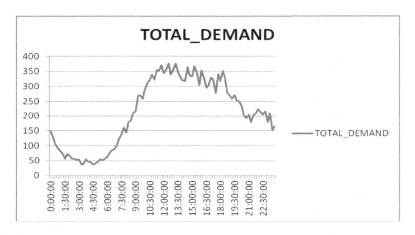

FIGURE 5.1 I=95 Express 24-Hour Demand

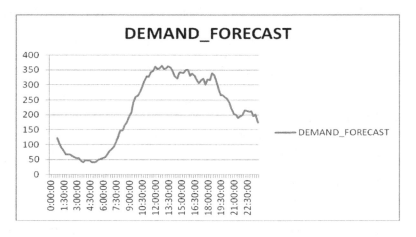

FIGURE 5.2 WMA Demand

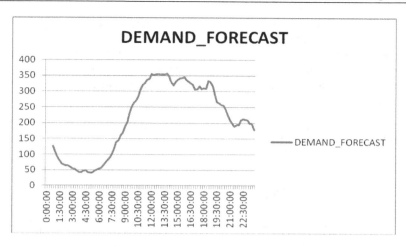

FIGURE 5.3   EWMA Forecast.

noted on the x-axis. Thus, for a 24-hour period there are 24*4 = 96 periods for which demand is recorded.

Demand volumes are plotted on the y-axis of the figure. Figure 5.2 shows the WMA forecast.

The EWMA forecast for the same day and station is generated and shown in Figure 5.3.

A GARCH (1,1) model was calibrated for this data, and the parameter estimates are shown here in Table 5.1:

TABLE 5.1   GARCH Model Fit.

GARCH(1,1)		GOODNESS-OF-FIT		
PARAM	VALUE	LLF	AIC	CHECK
μ	0.00	64.38	−122.75	1
α_0	0.00			
α_1	0.17			
β_1	0.82			
β_1	0.98			

Residuals (standardized) Analysis							
	AVG	STDEV	SKEW	KURTOSIS	NOISE?	NORMAL?	ARCH?
	−0.04	0.97	−0.33	−0.05	TRUE	TRUE	FALSE
Target	0.00	1.00	0.00	0.00			
SIG?	FALSE	FALSE	FALSE	FALSE			

Volatility forecasts were calculated for a 24-period time horizon and are shown below in Table 5.2.

These results show that GARCH is applicable to forecasting demand from time-series transportation data. This represents a new approach to transportation model forecasting and can be applied to improve forecast accuracy over other methods.

TABLE 5.2 Volatility Forecast

STEP	MEAN	STD	TS	UL	LL
1	0.004405	22026.46579	22026.46579	683.3545	−683.34569
2	0.004405	20097.00505	21083.81853	666.0829	−666.07408
3	0.004405	18363.36968	20217.71627	649.5128	−649.50394
4	0.004405	16803.42508	19420.49979	633.6092	−633.60036
5	0.004405	15397.76547	18685.36555	618.3392	−618.33043
6	0.004405	14129.35198	18006.26288	603.672	−603.66317
7	0.004405	12983.20182	17377.80418	589.5782	−589.56934
8	0.004405	11946.12103	16795.18664	576.0302	−576.0214
9	0.004405	11006.47416	16254.12367	563.0021	−562.99332
10	0.004405	10153.98574	15750.78491	550.4694	−550.46057
11	0.004405	9379.568638	15281.7436	538.4087	−538.39993
12	0.004405	8675.175636	14843.93039	526.7983	−526.7895
13	0.004405	8033.670754	14434.5928	515.6174	−515.60856
14	0.004405	7448.717656	14051.25946	504.8463	−504.8375
15	0.004405	6914.68265	13691.70876	494.4666	−494.45778
16	0.004405	6426.550294	13353.94108	484.4607	−484.45185
17	0.004405	5979.849854	13036.15444	474.8119	−474.80307
18	0.004405	5570.591139	12736.72287	465.5045	−465.49568
19	0.004405	5195.208444	12454.1774	456.5236	−456.51475
20	0.004405	4850.511504	12187.18922	447.8549	−447.84612
21	0.004405	4533.64253	11934.55474	439.4852	−439.47636
22	0.004405	4242.038528	11695.18243	431.4015	−431.39269
23	0.004405	3973.398207	11468.08107	423.5918	−423.58303
24	0.004405	3725.652877	11252.3494	416.0447	−416.03586

I-95 Lessons Learned

6

In this research we have completed a review and analysis of a broad range of topics related to dynamic pricing in the United States. In the initial phase of the research we embark on a review of congestion pricing implementations in the United States and internationally. We then look into different types of congestion pricing strategies (HOT Express lanes, Cordon pricing, etc.) which have been utilized across these implementations.

The main research goals of this research are threefold:

(1) Develop an efficient tolling algorithm for reducing congestion on I-95 HOT lanes
(2) Evaluate and compare the performance of three stochastic NLP algorithms (FDSA, GA, and SA to solving the dynamic programming problem
(3) Provide an efficient forecast of total demand D, which can be used to plan for Phase 2 of 95 Express implementation

The first goal of developing an efficient tolling algorithm was completed by adopting a system-optimal approach to the problem. By utilizing the well-known Akcelik travel time equation, we were able to formulate the dynamic programming problem into an NLP with the objective of minimizing total travel time along the corridor.

The Cell Transmission Model (see Daganzo, 1993) was applied to construct and analyze the flows into and out of the HOT lanes and GP lanes for each time period. Because this model is robust we can utilize large cell lengths approaching that of the HOT lane in Phase I of the study. The approach to the problem is to determine the optimal flows along the HOT and GP lanes for each Demand D, in time period k which will minimize the total travel time and thus provide a system-optimal solution.

The tolling algorithm thus developed uses these flows to determine congested travel time along the HOT lanes. These travel times are then supplied to Hobeika's (see Lee and Hobeika, 2003) toll equation to calculate the optimal toll for this demand D.

DOI: 10.1201/9781003358701-6

Our methodology accounts for the actual dynamic flows on the HOT lanes and uses these deviations from optimal flows to adjust the toll using price elasticity. Thus, for example, when we solve our NLP problem we may obtain a value of 1200 veh/hr for optimal flow along the HOT lane. However, the actual flow may be 1500 veh/hr. We therefore use price elasticity to reduce the actual (1500 veh/hr) to the optimal (1200 veh/hr). The tolling algorithm first solves for the base toll using the optimal values and then also adjusts the toll based on the deviation of the actual flow from optimal using an appropriate amount determined by price elasticity, which will reduce the flow to optimal values for the next time period. We have outlined this tolling algorithm and preliminary analysis with that due to the fact that we began with an initial toll of \$0.50 (as compared to current \$0.25 initial toll). Our proposed toll structure is comparable and slightly lower than exists on I-95 Express.

The second goal of this research is to evaluate and compare the performance of three NLP algorithms to solving the dynamic programming problem. The three NLP algorithms we chose to evaluate were the Finite Difference Stochastic Approximation (FDSA), Simulated Annealing (SA), and Genetic Algorithm (GA). For this purpose, we employed MATLAB software to generate a coded version of the NLP problem and then for fixed, known levels of demand and density applied the three algorithms separately to produce the optimal solution. The optimal flows generated by each algorithm were then tabulated and the travel time was recorded.

These travel times were analyzed in Minitab statistical software. The analysis performed consisted of a two-way ANOVA to determine if there exists any significant difference in the algorithm's performance in minimizing travel time. Thus, the response variable of interest was travel time, and the factors (explanatory variables) were (1) algorithm and (2) demand level.

The travel time values were obtained by simulating the traffic flow along the network in VISSIM software. Having performed multiple simulation runs for each demand level varying from D=2000 vph to D=16000 vph, we sum the individual values to obtain the total travel time for all the flow during the period. These totals are then analyzed in ANOVA using the 2-way ANOVA method.

The results of the ANOVA analysis show that that there is significant difference in performance between the algorithms at certain level of demand. The data analysis provide evidence that with this problem formulation and simulated travel time, at demand level 13,000 vph the FDSA and GA algorithms perform significantly better than the SA Algorithm. There was no apparent significant difference between the FDSA and GA and SA algorithms observed above 14,000 vph.

The results of the two-way ANOVA show that for demand volumes less than capacity the three algorithms show no significant difference in performance. However, at demand volumes slightly exceeding capacity (of 12,000 vph) the SA algorithm produces travel times that are significantly higher than the other two algorithms.

The third goal of this research is to accurately forecast demand flows on the corridor on an hourly basis. For this purpose we applied a new technique, which is based on the economic forecasting of dynamic pricing in stock markets.

The data for the month of October 2012 was utilized to perform this analysis. A GARCH volatility model was calibrated using the 15-minute demand data from the STEWARD database. We selected station data collected from district 6 on I-95 HOT lane northbound to perform the analysis. This data was used to calibrate a GARCH (1,1) model, and volatility (%change) for intraday, hourly demand was forecasted in EXCEL.

The results show that econometric forecasting techniques can be successfully applied to transportation problems to identify and calibrate accurate models.

In summary, this research brings a fresh approach to the topic of dynamic pricing. It demonstrates how system-optimal methodology can be applied to minimize travel time along the 95 HOT lanes and reduce congestion with an efficient tolling algorithm. This methodology is valuable for practitioners since it is highly portable. It can easily be applied to congested urban areas such as Washington D.C., which already utilizes HOT lanes, and other large cities such as Atlanta and Orlando. It also represents a new, improved approach to pricing on I-95 Express HOT lanes. We believe this tolling methodology is efficient and can be quickly implemented to current transportation HOT infrastructures and the I-95 phase 2, which will be completed sometime in 2014.

Congestion Charge Case Studies

7

7.1 THE LONDON CONGESTION CHARGE

The City of London has instituted a charge for drivers to access the downtown area or the central London district. This charge will apply to all vehicles driving to the downtown area between the hours of 7:00 a.m. to 10:00 p.m. at night, except for certain exempt vehicles such as electric vehicles.

The London congestion charge was first instituted in 2003 and was intended to reduce traffic flow and pollution in the city center of London. The congestion charge is a fee of 15 pounds, which is charged on a daily basis for vehicles to access the downtown central business district of London. Certain vehicles are exempt from this fee, and also certain residents that live within the area of the congestion charge zone are exempted from paying this fee.

In addition to the congestion zone, there is also the Ultra-Low Emission Zone (ULEZ), which has a fee of 12.50 pounds for cars. All diesel vehicles more than four years and all gasoline cars manufactured before 2006 are required to pay this fee in the ULEZ. The pollution level in London is seen to cause tens of thousands of people to become seriously ill and is linked to deaths of many others from diseases such as asthma. Pollution costs 3.7 B pounds per year according to the mayor of London. The Transport from London

Tries to move people for polluting vehicles to cleaner vehicles such as electric vehicles and to public transportation such as Subway systems and buses.

The London congestion charge comes with its own controversy, and it has its own pros and its own cons.

DOI: 10.1201/9781003358701-7

Business owners in the central district of London have complained that the congestion charge has reduced traffic and the number of customers that come into their stores. This in turn reduced the amount of annual revenue that they were able to earn from walk-in or drive-through traffic on a daily basis. In addition, many residents just see the congestion charge as a tax on the poor and a money grab by their governing representatives; they often complain that although there have been some benefits for some people by and large, the congestion charge has not worked and has not achieved the objectives that it was initially supposed to have achieved.

7.1.1 Brief History of Congestion Pricing

Congestion pricing was first recommended by Nobel economist William Vickery and the 1950s. Vickery was a professor of economics who introduced the use of charging a fee to change or modify the behavior of the public use of transportation facilities.

The city state of Singapore was one of the first major urban regions to introduce congestion pricing successfully. In the 1970s, Singapore began charging a fee for drivers to enter the downtown city area during certain hours rare there was peak traffic.

Following up on the successful implementation of congestion pricing in Singapore, other urban regions also started to implement congestion pricing schemes.

The City of London introduced congestion pricing and their downtown central business district in 2003. Initially, this pricing schema was in effect between the hours of 7:00 a.m and 6:00 p.m. but later on changed till 7:00 a.m. to 10:00 p.m., and most recently the pricing also increased to 15 pounds daily. The congestion pricing zone and the fee that has been charged hazmat with a lot of negative sentiment and many see it as just another tax and disproportionately affecting the poor segment off the population which are forced to live and work in the congestion zone.

Other cities such as Stockholm Sweden have also implemented their own congestion pricing scheme for their downtown regions and in their case what was initially our poor reception by their residents rare only 25% of the population agreed or so the charge as positive later on found that over 70% of the population agreed with the charge after seeing the effect that it had not only on congestion but also on reducing pollution in the urban region.

New York City has also made plans to start pricing for traffic and they're Manhattan downtown area all the way up to 60th St. Due to the COVID-19 pandemic, this plan was put on hold but in the recent months of 2022 new energy has been put In Focus on implementing the congestion scheme shortly.

William Vickery received the Nobel Prize in economics for his pioneering work in the economics of incentives. He was one of the first researchers to present and introduce the concept of congestion pricing on the roads in the United States. His ideas in congestion pricing stemmed from the observation that in populated cities such as New York City space was a valuable commodity and the use of this space as well as transportation space on roadways should only be available at a cost. Vickery also applied his concepts to the market for electricity and proposed pricing mechanisms for consumers using electric power from the grid. Another pioneering work which Vickery presented was the application of game theory from mathematics to the operation of auctions. Vickery also developed models for taxation in the Japanese economy After World War Two.

Congestion pricing extends beyond the concept of charging a fee for using a certain area or region in a city. In a more general sense congestion pricing refers to a charge which is assessed for using a facility such as a road network or communication network when the demand for usage exceeds the supply that is available. In this way the congestion fee can be used to reduce the demand so that the supply balances the available resource demand and preserves the quality of the experience of using the facility. One of the purposes of the congestion price is to ensure that delays are minimized on a network; so, for example, when a highway implements an express lane with a congestion price, the objective is to reduce the traffic on that lane so that the demand is less than or equal to the available supply capacity for transportation on that lane.

7.1.2 What Causes Traffic Congestion?

Traffic congestion is caused whenever the demand for the traffic facility exceeds the supply or capacity of the roadway. This can be due to a number of reasons or number of causes. People who live in urban regions or large cities experience traffic congestion on a daily basis due to repeated or recurring daily demand. This demand exists due to large numbers of drivers who must drive to work and who must be on the road at the same time. Thus, in the morning hours between 7:00 a.m.and 9:00 a.m. there is a spike in traffic volume on the roads in urban locations; this is due to commuters traveling to work typically in a downtown area of the city. In the same way between the hours of 4:00 p.m. and 6:00 p.m. there is a great increase in demand for using the same roadway in the reverse direction, and again this results in the demand exceeding the capacity of the roadway, which causes traffic congestion. This spike in demand occurs on a daily basis and thus it's a recurring repeated source of traffic congestion.

Traffic congestion can also be caused by accidents on the roadway, which blocks several lanes of traffic and therefore significantly reduces the supply and the capacity of the roadway. Thus, even though there is no increase in demand for the road facility, there is a significant decrease in supply, and therefore this also results in traffic congestion. This type of traffic congestion is non-recurring and it's not necessarily on a daily basis.

Another cause of traffic congestion is construction, which typically blocks off a lane or multiple lanes of the highway, which again reduces the supply of roadway for vehicles to flow.

In the case of bad weather there can be a pile up of vehicles due to snow, heavy rain, or even hail. Because of these inclement conditions, large numbers of vehicles can be forced to drive in a smaller area, and this also causes congestion.

7.1.3 Implementation

7.1.3.1 Ultra-Low Emission Zone

The ultra-low emission zone (ULEZ) is a zone in the City of London that covers the same area as a congestion charge zone. The main purpose of this zone is to focus on pollution from certain vehicles and to minimize this pollution in the central business district of London. Unlike the congestion charge, the ultra-low emission zone will be in effect 24 hours a day, seven days a week, except for Christmas Day. The ULEZ charge is £12.50, and this is in addition to the congestion charge fee.

There are certain types of vehicles that will have to pay this fee if they drive through the ultra-low emission zone in the central district. The vehicles that will be assessed this fee are the following:

1 Diesel cars registered before the year 2015
2. Gasoline cars registered before the year 2006
3. Motorbikes registered before the year 2007
4. Trucks and buses that do not meet the Euro 6 standard.

The Euro standards for vehicles started in 1992 and prescribe maximum limits on the amount of harmful pollutants that can be emitted from diesel and gasoline vehicles. The purpose of the Euro standards was to reduce the nitrogen oxide (NOx) and carbon compounds released into the environment due to vehicular combustion of these fuels. The standard progressed from Euro 1, Euro 2, Euro 3, Euro 4, Euro 5, and Euro 6. In the year 2015, the Euro 6 standard was set to be a maximum of 80 mg/km for diesel engines and 60 mg/km for gasoline engines as shown in Table 7.1 and 7.2.

TABLE 7.1   Euro Standards for Diesel

EURO STANDARD (DIESEL)	DATE	NOX	CO	PM
Euro 1	1992	–	2.72	0.14
Euro 2	1996	–	1.0	0.08
Euro 3	2000	0.50	0.64	0.05
Euro 4	2005	0.25	0.50	0.025
Euro 5	2009	0.18	0.50	0.005
Euro 6	2014	0.08	0.50	0.005

TABLE 7.2   Euro Standards for Gasoline

EURO STANDARD GASOLINE	DATE	NOX	CO	PM
Euro 1	1992	–	2.72	–
Euro 2	1996	–	2.2	–
Euro 3	2000	0.15	2.3	–
Euro 4	2005	0.08	1.0	–
Euro 5	2009	0.06	1.0	0.005
Euro 6	2014	0.06	1.0	0.005

The recent emissions scandal at car manufacturer Volkswagen shows how important the emissions problem is and also how the current tests for emissions can be circumvented to produce results that appear to comply but which are not compliant to the regulations.

The ultra-low emission zone in London charges a fee of £12.50 to drive a vehicle that does not meet the standards through the zone. An additional fine of £160 is charged if that fee is not paid; however, if the offender pays the fee in less than 14 days only £80 will be charged through the Transport for London website (Tfl.gov.uk).

According to Transport for London, most vehicles that were produced after the year 2005 will follow the ultra-low mission zone standard of Euro 4 emission standards for gasoline engines.

According to the website www.gov.uk/plug-in-vehicle-grants, there are grants provided by the government of the United Kingdom to assist with the purchase of compliant vehicles such as vans, light trucks, wheelchair-accessible vehicles, and motorcycles or mopeds. In addition, there is a scrappage scheme that allows for a grant to replace older these diesel vehicles and to purchase one that is compliant with the ultra-low emission zone.

7.1.4 Operation

Initially, when the congestion charge was implemented in 2003, the traffic driving through the central business district was reduced by about 15%. As time went by, the results have been mixed. with some sources reporting that traffic increased during later years.

Since the start of charging the congestion charge in London's central business district, there have been several adjustments in the location of the areas to be charged such as an extension to the western side of the district there has also been a change in hours from the initial 7:00 a.m.to 6:00 p.m. charging. And in addition there have been changes in the standards that the vehicles must meet in order not to be charged the emission fee.

The Transport for London organization has continuously monitored the traffic flow through the congestion zone and prior to implementing the charge in 2003 has performed a study of the traffic flow before the charges were implemented. In this way they were able to compare the impacts of the congestion fee after implementing the fee in the central business district to determine the changes in traffic flow as well as emissions to perform a before- and after-impact study of the congestion charge.

In their study, Transport for London asserts that between the years 2002 and 2007 there has been a reduction in traffic volume of approximately 16% for all vehicle types and 21% for vehicles with four or more wheels. Their data also shows that they have maintained this reduction over that period of time.

Their data shows a 36% reduction in traffic volume for cars, a 13% reduction in traffic volume for vans, a 7% increase in taxi volume and a 31% increase in traffic volume for buses.

The data was collected using both automatic traffic counters as well as periodic manual counting, and the city installed a total of 18 automated traffic counters in the central district and an additional two automated counters forward a western extension.

From the above table, we also see a marked increase in the use of pedal cycles of 66% that entered the central congestion charging zone during the year 2007 over 2006. Buses and coaches as well as licensed taxis also increased in numbers during that period of time. Charging hours were changed after the February 9, 2007, from 18:30 to 18:00 for the ending time.

TABLE 7.3 Year-on-Year Changes to Traffic Entering the Central London Charging Zone During Charging Hours

	2003 VS 2002	2004 VS 2003	2005 VS 2004	2006 VS 2005	2007 VS 2006	2007 VS 2002
All vehicles	−14%	0%	−2%	0%	0%	−16%
Four or more wheels	−18%	−1%	−2%	−1%	0%	−21%
Potentially chargeable	−27%	−1%	−3%	0%	1%	−29%
- Cars and minicabs	−33%	−1%	−3%	−1%	0%	−36%
- Vans	−11%	−1%	−4%	2%	1%	−13%
Non chargeable	17%	1%	−1%	−1%	−1%	15%
- Licensed taxis	17%	−1%	1%	−3%	−5%	7%
- Buses and coaches	23%	8%	−4%	−3%	5%	31%
- Powered two-wheelers	13%	−2%	−9%	0%	−3%	−3%
Pedal cycles	20%	8%	7%	7%	12%	66%

7.1.5 Boundary

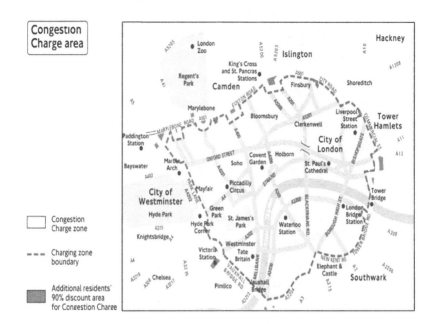

FIGURE 7.1   London Congestion Zone

(Source: Transport for London. tfl.com.uk)

7.2 THE STOCKHOLM CONGESTION TAX

The Stockholm congestion tax was implemented on August 1, 2007. It was implemented after a trial period of seven months in 2006 when the system was tested and then a referendum was held, and the congestion tax was voted on. The congestion fee amount varied during the day from a base price of 10 SEK to 20 SEK depending on if it was peak hour or not. This price was valid until December 2015. After December 2015, the price was increased for the new year of January 2016.

Tables 7.4 and 7.5 show the hourly rates during the charging period before the increase and after the price increase.

There are 18 entry points into the Stockholm city center, and the toll gantries are set up to read both the front and rear license plate numbers using

TABLE 7.4   Stockholm Congestion Tax Charging (Before Price Increase 2016)

TIME OF DAY	FEE (K)
06:30–06:59	10
07:00–07:29	15
07:30–08:29	20
08:30–08:59	15
09:00–15:29	10
15:30–15:59	15
16:00–17:29	20
17:30–17:59	15
18:00–18:29	10

TABLE 7.5   The Congestion Fee After the Price Increase (January 2016)

TIME OF DAY	FEE (K)
06:30–06:59	15
07:00–07:29	25
07:30–08:29	35
08:30–08:59	25
09:00–09:29	15
09:30–14:59	11
15:00–15:29	15
15:30–15:59	25
16:00–17:29	35
17:30–17:59	25
18:00–18:29	15
18:30–06:29	0

automatic number plate recognition. Once the system reads the plate, the congestion tax is deducted from the account holder's account.

Any car owner may set up an account in a local convenience store such as Pressbyran, by purchasing an onboard unit (transponder) and signing the contract. Anyone may also sign up online on the government website and receive a transponder onboard unit by mail.

The congestion fee is assessed both when entering and leaving the congestion zone through any of the 18 entry/exit points.

7.2.1 Operation

The congestion tax is paid after the driver has been recorded by either the Automatic Number Plate Recognition (ANPR) or by detecting the transponder unit and deducting the fee directly from the associated account.

The technical infrastructure of the congestion tax collection system consists of overhead gantries, which allows for uninterrupted flow of traffic. There are no toll booths, and vehicles do not stop when entering the charging zone or when leaving the zone to pay the congestion tax. The components of the system include

(1) Laser detectors
(2) Cameras (for front and rear number plates)
(3) Transceivers

The operation of the overhead toll gantry follows the steps:

1. Vehicle breaks first laser beam detector
2. Aerial transceiver sends signal to the onboard unit
3. The onboard unit responds by sending account information to the transceiver
4. Vehicle then breaks the next laser beam
5. The aerial cameras then take a picture of the front number plates only
6. When the vehicle leaves the range of the second laser beam, the aerial cameras take a picture of the rear number plate

After the vehicle leaves the congestion zone, the congestion tax is calculated, and the tax is deducted from the owner's account. The amount of the congestion tax is available online usually within a 24-hour period. The payment can be made at a convenience store such as Pressbyra or via direct debit from the onboard system account.

The congestion tax was increased in 2016. After this increase, there was a marked decrease in traffic volumes going into the city center.

Several studies have been done comparing the price elasticities of the congestion zone before the price increase and after the price increase. Studies have also been done to compare the costs and benefits of the congestion charging scheme at initial implementation and as time progressed to the increase in prices during the year 2016.

Borjesson (2017) estimates a reduction in traffic volumes from 30,898 to 29,315 and a price elasticity constant of −0.28 in response to the price increases in 2016.

We also see that the price elasticity varies across different modes of transport, and whether it is a private vehicle versus a commercial transportation vehicle such as trucks (freight) that are relatively price insensitive as they are required to make deliveries regardless of the congestion fee.

The congestion fee in Stockholm initially received a tepid response from the population, with only approximately 30–40% of commuters in favor of the congestion charge. However, after the charge was introduced, between 2006 and 2015 the popularity of the charge increased up to as much as 70% when the urban population was able to see the reduction in congestion and accidents due to the charge being imposed.

After 2016, when the congestion fee increased, there was a decrease in the number of people who were in favor of the charge since the traffic volumes decreased only moderately, and the perception was that the charge was being used more for revenue generation purposes than to really decrease congestion and pollution.

7.2.2 Impacts and Lessons Learned (Stockholm Congestion Charge)

The investment in the Stockholm congestion price system was initially approximately 200 M euros (Eliasson, 2009). According to Borjesson (2017), yearly revenues from the system varied between 70 M (euros) in 2008 to 140 M euros in 2016 while yearly operating costs were between 9.6 M euros to 22 M euros during the same time period.

The Stockholm congestion charging system utilizes both a transponder onboard the vehicle as well as the ability for Automatic Number Plate Recognition (ANPR). Due to the Automatic Number Plate Recognition system having a low (60–70%) effectiveness, the onboard transponders were also used to identify the vehicle and deduct funds from the user's account. However, due to legal reasons, an onboard transponder alone was not allowed for the city of Stockholm's congestion tax.

Studies done at the Royal Institute of Technology Center for Transportation Institute in Stockholm (2010) show that the transponder was not absolutely necessary. In fact, when the Automatic Number Plate Recognition (ANPR) system is paired with a high-resolution camera, the accuracy of the vehicle ID is greatly improved and is sufficiently reliable for accurate identification and charging. The study also points out that transponders do have the added advantage of being portable from vehicle to vehicle,[1] which means the owner of several vehicles may have a single account and pay one congestion tax for all vehicles.

The congestion tax has resulted in a 23% decline in traffic congestion in Stockholm. The decline has been sustained over the years and reduced the emissions and traffic accidents in the congestion zone.

As mentioned before, the initial reduction in traffic and price elasticity is not replicated with subsequent price increases and public support can drop with additional price increases.

For trucks and other commercial vehicular traffic, they are relatively insensitive to price due to the requirements for their travel into the congestion zone.

The data shows that operation costs and investments decline over the time period since the introduction of the congestion tax. Another benefit of the congestion fee is the increased usage of public transport and non-motorized vehicles. Revenue generated from the congestion tax is re-distributed into expanding public transport.

Other pertinent facts about the congestion tax include the following (see www.visitsweden.com):

1. The tax has also been charged in Gothenburg since 2013 with a different rate schedule.
2. The congestion tax is not charged on Saturdays, Sundays, holidays, or in the month of July.
3. Negative impacts on the economy were minimal with reduced CO_2 emissions.
4. Emergency vehicles, mopeds and motorbikes are exempt from the congestion charge.
5. If a vehicle drives under multiple access points to the congestion zone during the same day, only the highest fee is charged to that account.

IBM was selected as the main contractor to develop design and install the technical infrastructure for the congestion charging system in Stockholm. One of the lessons learned was that the Automatic Number Plate Recognition system can be improved by using video analytics, and this was implemented for reading license plates with higher accuracy.

As of January 1, 2020, several adjustments were made to the congestion tax in Stockholm; these include seasonal adjustments to include a peak season and off-peak season, charging the tax during the first five days in July, and taxing some days prior to a holiday.

7.2.3 Seasonal Adjustments

Based on the traffic patterns in Stockholm throughout the year, it was determined that late spring, summer, and early autumn experience higher traffic flows than that of the rest of the year, and this period was designated as "peak season." The rest of the year is designated as "off-peak season."

During peak season the maximum daily congestion charge is higher than the maximum charge during off-peak season.

7.2.4 Boundary

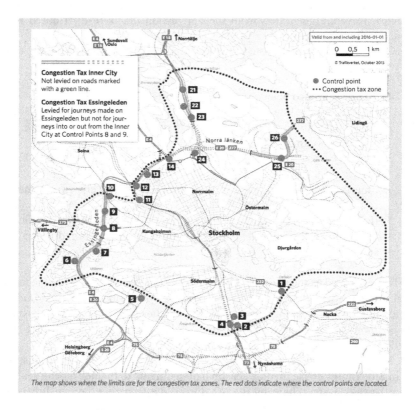

The map shows where the limits are for the congestion tax zones. The red dots indicate where the control points are located.

FIGURE 7.2   Stockholm Congestion Zone Map

(Source: maps.stockholm.com)

7.3 SINGAPORE CONGESTION FEE

Singapore exists as a city-state, which means it essentially has no capital as the entire country consists of the city. Singapore has a very small area and high population density. It has a population of approximately 4.5 million people and roughly 250 square miles. This high population density results in traffic and transportation problems, which led to Singapore developing transport demand management schemes in 1975 to help reduce the traffics congestion it experienced particularly in the downtown "restricted zone" (RZ).

In 1975, Singapore embarked on its Area Licensing Scheme, where vehicle owners traveling into the downtown area restricted zone (RZ) were required to buy a license in order to enter the zone through one of the 28 entry points. This license was sold at retail outlets and other locations for the equivalent of US$1.30.

The vehicle owners then were required to display the paper license in their vehicle when they drove in the restricted zone or risk a hefty fine.

Singapore has also made it very expensive for private citizens to purchase automobiles in an effort to reduce the number of vehicles and congestion on the roads and highways.

There are certain exemptions to the congestion fee for entering the restricted zone. For example, buses, emergency vehicles, high-occupancy vehicles, and motorcycles were exempted from having to pay the congestion fee when entering the restricted zone.

The congestion fee was initially charged during the hours from 7:30 a.m. to 9:30 a.m., which was the peak rush hour when most employees who worked in the restricted zone were traveling to work.

Between 1975 and 1988 a number of changes were made to the charging scheme. For one, the charging end time was extended from 9:30 a.m. to 10:15 a.m.

The rates have been adjusted several times, and eventually charging times were extended throughout the daytime. In addition, some of the previously exempt vehicles such as taxis lost their exempt status and were charged the congestion fee.

7.3.1 Singapore Congestion Charge-Operation

The Singapore congestion charge transitioned to Electronic Road Pricing in 1998. This allowed for the commuters and owners to pay the fee without using a paper license. With Electronic Road pricing, the drivers also do not have to stop at any point on the highway or in the restricted zone in the central business district to pay the fee, and the infrastructure is capable of reading an onboard unit and collecting the fee at high rate of speed.

Each traveler in the central business district and at multiple points throughout the city must possess an onboard unit, which is a transponder capable of being read by overhead gantries. A "smart" card is used with the transponder, which has account information as well as other vehicle-type identification information stored on it. The smart card information is transmitted to the overhead gantries when the vehicle travels past the charging point, and the charge is then deducted from the user's account at that time. If the transponder does not work or if the vehicle does not have a transponder, the camera on the overhead gantries takes a picture of the license plates, and a bill is sent to the owner of the vehicle.

Fines are assessed for various types of violations, and bills are sent to the owner accordingly.

Singapore has experienced a high rate of compliance, and it is reported that less than 1% have "cheated" the system and not paid the congestion fees.

7.3.2 Singapore Congestion Fees-Impacts

After the introduction of the Area Licensing Scheme in 1975, there were noticeable changes to the traffic patterns around the restricted zone (RZ). There was a significant decrease in vehicular traffic from approximately 74,000–41,000 and a corresponding decline in cars of more than 70% (see ops.fhwa.dot.gov). Simultaneously, there was an increase in the usage of buses and rapid transit as public transport rose from 33% to 69% between 1976 and 1983.

Other changes in travel patterns were observed, such as an increase in traffic around the central business district, which in turn lowered the average speeds and delays in these alternate routes.

Speeds of travel inside the restricted zone increased by at least 20% and averaged 30–35 Km/hr. Initially, when the ALS was first introduced, there was a surge in traffic at the end of the charging period (9:30 a.m.), and so the charging time was extended to 10:15 a.m. to allow for a smoother transition from peak traffic to lower, less congested levels. Along with the reduction in traffic, there was also a reduction in pollution levels and an increase in pedestrian and bike usage.

The reaction to the implementation of the congestion charge was mixed. For the most part, people traveling from outside the zone to work inside the zone would respond negatively or neutral to the congestion charge. However, for those people who lived in the zone, the response was more positive and accepting because they experienced fewer accidents and less congestion around their place of residence.

With less traffic, there is less noise pollution, less CO_2 emissions and fewer accidents, the people living in the restricted zone experience a higher quality of life and more conducive environment for business to flourish.

When the congestion pricing program began, there were only 28 toll gantries that were placed at the entry points to the central business district (RZ). As the Electronic Road pricing was implemented, there were over 80 points in the city-state, where toll gantries were placed on the highways.

The impact of expanding the number of tolling points has been significant economically, and with more revenues generated from the charging scheme, Singapore has been able to expand its public transportation network and increase ridership on its buses and mass transit system.

The congestion price is dependent on the time of day, the type of vehicle, and the actual road conditions and congestion on the highway. The revenues generated range up to six times the annual operating costs, which allows for an investment in better safety and efficiency in the transportation services.

Besides the reduction in traffic and increase in speeds in the restricted zone (RZ), the Land Transport Authority in Singapore reports a 15% decline in CO_2 emissions and other pollutants when compared to levels prior to the implementation of the charging system.

As motorized vehicle trips have reduced in the restricted zone, there has been an increase in the number of trips by bicycle and pedestrian modes of transport, which has resulted in significant health benefits when combined with the lower pollution levels in the central business district.

Other initiatives have also played a role in reducing the congestion and polluting effects of vehicular traffic in Singapore. The city has increased (doubled) the cost to park in the downtown area, thus deterring the presence of cars from entering the city unless absolutely necessary. There have been a number of taxes and fees on the purchase of an automobile in Singapore, and the government requires new car buyers to purchase a Certificate of Entitlement in order to buy a car in Singapore. This certificate costs in excess of $40,000 in addition to the cost of the car and other taxes. New certificates for car purchases were also capped in 2018.

Although Singapore's population has grown tremendously since the start of the ALS congestion pricing, the number of vehicles and the congestion have been kept at a relatively low level. This is one measure of success for the charging system and demonstrates it is well designed and very effective.

Singapore's congestion charging scheme is operated between the hours of 7 a.m. and 8 p.m. and the fares range from $0 to $3.00 per charge. The charges are suspended on certain weekend days and holidays. Due to an overall perception of fairness of the charging fees and due to the positive impacts on reducing congestion and pollution, many practitioners and citizens see it as a success.

With several other major international cities such as New York City planning their own congestion pricing charge, it is well worth the time to look into what Singapore has done over the last 40 years to be successful in this area.

7.3.3 Economic Impacts

Despite the introduction of the Area Licensing Scheme (ALS) fee and subsequently the ERP programs, there does not appear to be major negative economic impacts on the businesses in the restricted zone or over the city-state in general. Most businesses in the restricted zone have shown no negative impacts to the sales of products or services. The only exception reported were some minor to moderate impacts when the charging was expanded from only 7:00 to 9:00 in the morning to include afternoon charging as well (www.ops. fhwa.dot.gov). This seems to have been a temporary phenomenon as customers adjusted to the new time structure of the ALS charging system.

One of the key factors in the success of the Singapore charging system is the ability of the administration to successfully shift traffic from automobiles to mass transit. The heavy investments into providing clean, safe reliable public transportation made it easier for citizens to change their mode of traveling into the restricted zone and thus resulted in no loss of business for the central business district.

7.3.4 Boundary

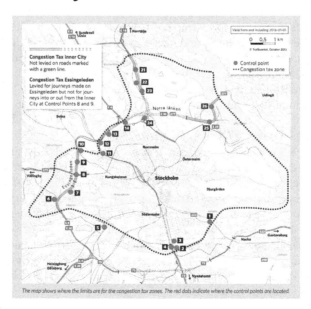

FIGURE 7.3   Map of Singapore ERP

(Source: itf-oecd.org)

7.4 MILAN CONGESTION CHARGE

The city of Milan is a large metropolitan urban region in Italy. It is the second-largest city with a population of over 1 million residents. The larger metropolitan region of Milan (Greater metropolitan area), including the outlying areas, has a population of over 8 million people.

Milan has experienced congestion within its downtown are for several years and has suffered from pollution problems as well.

Because of these congestion issues, Milan city government developed "Area C," which is a region in its downtown area and where a congestion fee was implemenmted in 2012.

Area C is the successor to the prior system called Ecopass but occupies the same area. The Ecopass program fee applied only to vehicles with high polluting emissions that did not meet certain Euro pollution standards. Diesel vehicles that did not meet diesel Euro 3 standards and gasoline vehicles that did not meet Euro 0 standards cannot enter the restricted zone.

The congestion charge ranges from 2 euros to 5 euros depending on whether the vehicle is resident or non-resident, and whether the vehicle is a commercial or non-commercial vehicle.

The congestion fee is charged between the hours from 7:30 a.m. to 7:30 p.m., with temporary changes due to the COVID-19 pandemic, when the charging time was modified.

Vehicles that are exempt from the congestion fee include electric vehicle, public transit vehicles, motorcycles, emergency, and police vehicles.

7.4.1 Operation

The Milan Area "C" congestion charge utilizes 43 entry points where cameras are used to detect commuters' license plates with Automatic Number Plate Recognition (ANPR). Residents who live within the restricted zone get 40 free passes each year and are only charged 2 euros starting on the 41st access. The city administration realized the shortcomings of the Ecopass charging scheme and made the necessary changes in operation of the system.

The Area "C" charging scheme was introduced in 2012 after a pilot program and referendum. The program received majority support from its population of users, when in excess of 75% voted in favor of the scheme. This figure is greater than the popular approval for the Stockholm congestion tax.

7.4.2 Impacts

The charging scheme has resulted in a shift in mode for traffic entering the Area "C." Some sources report up to 20% increase in bicycle usage through the city center, and Milan has invested in a cycle network that consists of concentric paths around the center that is intended for bicycle traffic.

In a similar fashion, Milan's Area "C" experienced significant shifts from private vehicle travel to public transit such as buses and trams. Reduction in traffic and congestion was achieved on the order of 25–30% less than before the congestion fee was introduced and a corresponding reduction in accidents of approximately the same number.

Pollution levels and smog were also impacted in a positive way. Carbon dioxide levels have been reported to be reduced by 30–40% when compared to levels before the implementation of the congestion fee, and particulate matter (smog) has been reduced by 8–12%.

The reduction in traffic levels resulted in higher average speeds for the buses and mass transit, translating into higher satisfaction levels for commuters on public transportation.

The congestion charge is active on weekdays but not on weekends (Friday, Saturday, Sunday) or on holidays.

The goal of the congestion charge in Milan is not only to reduce traffic but also to reduce pollution. Milan has been known to have one of the highest percentages of car ownership in Europe, and with this high rate of individual vehicle ownership comes a high rate of pollution in the city and restricted downtown areas. The municipality of Milan has therefore instituted several bans on highly polluting vehicles depending on the type of vehicle.

Some of the bans concern vehicles carrying people, city resident vehicles, vehicles used for public transport and vehicles for freight. These bans rely on the Euro standards for gasoline and diesel engines and have exemptions for electric vehicles.

Milan's congestion scheme was instituted after Singapore (1975), London (2007), and Stockholm. As such, it had a lot of opportunity to learn from these implementations about what will work and how to do it effectively.

7.4.3 Payment

Payment of the congestion charges can be made through apps or by website. The TelePass app provides the option to pay Area C charges as well as other transport fees. If the Area C charge is not paid within seven days, there is a penalty fee that is added to the amount owed, and the owner of the vehicle is assessed this fee. Tickets to enter Area C can be purchased at newspaper agents, by website or by phone.

The Area C ticket must be activated by a specified time during the day of access or not later than the following day.

Area C has been successful in reducing noise pollution and plans to also introduce a Low Emission Zone similar to what exists in London (ULEZ) to further control pollution due to carbon dioxide and particulate matter.

Milan's congestion charge in Area C has generated tens of millions of euros annually (~28 M euros in 2016), while keeping operating costs low, and in 2015 the operating maintenance costs were along the order of 3.8 M euros.

The revenue generated by Area C congestion charges are re-invested into improving the infrastructure of the transportation system. For example, over 60% of revenue was spent on increasing the frequency of mass transit in one year. Another example is the new pedestrian and bike-sharing system developed to shift the mode of transportation for commuters in the city and reduce the vehicular traffic, which also reduces accidents and pollution.

Initially, third-party delivery services were against the Milan congestion fee, but after the implementation resulted in higher speeds and more free-flowing traffic, these services viewed it as a benefit for their service since it increased their business in the downtown area.

Other "Smart City-"related investments include

1. Expanding Car sharing services
2. Expanding Electric Bike Sharing Services
3. Expanding Scooter sharing

7.4.4 TelePass

TelePass is one of the ways to pay the fee for Area C; thus, we will describe its usage in this section.

TelePass is an App that is used to pay for various mobility services around the city of Milan. After the Area C option is activated in the TelePass App, the user will be able to pay only one time to access the restricted zone and travel into and out of Area C as many times in a day as needed. The Telepass app allows for the fees to be paid once per quarter. TelePass can be used to pay for a variety of services such as

1. Parking
2. Area C charges
3. Fuel
4. Vehicle Inspections
5. Public Transport
6. Ride Sharing Services
7. Ferry services

Telepass is an integrated mobility service that allows for toll payments in more than 10 European countries including France and Spain. It allows for payment in the Area C restricted zone on a bulk basis of 30 euros or even 60 euros to pay for multiple trips into the Milan congestion area.

Overall, Milan by implementing Area C congestion charge has been able to reduce traffic congestion significantly, reduce smog and other pollutants, which aligns with climate change goals, reduce vehicle accidents, increase pedestrian, and bicycle usage in the restricted zone and increase number and usage of buses and public transport. From this perspective, the congestion scheme has been a success. There have been social implications which have resulted in some concern among certain segments of the population which some may view as negative and the whole picture of how the system is operated and how it impacts different segments and regions of the state must be considered.

Milan has implemented Area B. This area lies outside of Area C and has been active since 2019. The area has strict regulation on emissions and set standards for all vehicles except electric vehicles, methane vehicles, and hybrid vehicles.

7.4.5 Boundary

FIGURE 7.4 Map of Milan Area "C"

(Source: maps-Milan.com)

NOTE

1. Börjesson, M., and I. Kristoffersson. "The Swedish Congestion Charges: Ten Years on: and Effects of Increasing Charging Levels," Working Papers in Transport Economics 2017:2, 2017. CTS—Centre for Transport Studies Stockholm (KTH and VTI).

Appendix A: Travel Time Measurements

The following table is the Minitab Worksheet for the 2-WAY ANOVA analysis.

DEMAND	ALGORITHM	TRAVEL TIME (MIN)
2000	SA	24931.39
2000	SA	25517.49
2000	SA	25770.87
2000	SA	27821.69
2000	SA	27106.9
2000	GA	26848.31
2000	GA	24877.09
2000	GA	26077.18
2000	GA	25001.81
2000	GA	25404.48
2000	FDSA	25445.45
2000	FDSA	25894.17
2000	FDSA	25083.78
2000	FDSA	25429.95
2000	FDSA	25210.42
3000	SA	38479.42
3000	SA	37876.31
3000	SA	38617.67
3000	SA	37984.78
3000	SA	37094.13
3000	GA	38482.09
3000	GA	37894.3
3000	GA	38061.4
3000	GA	37576.38
3000	GA	40941.97
3000	FDSA	38838.11
3000	FDSA	41308.13

DEMAND	ALGORITHM	TRAVEL TIME (MIN)
3000	FDSA	40018.76
3000	FDSA	39738.2
3000	FDSA	40637.89
4000	SA	53103.55
4000	SA	53103.55
4000	SA	53103.55
4000	SA	53136.97
4000	SA	53216.92
4000	GA	53117.18
4000	GA	54758.88
4000	GA	53258.2
4000	GA	52885.59
4000	GA	52849.22
4000	FDSA	53980.51
4000	FDSA	52801.22
4000	FDSA	52639.62
4000	FDSA	53210.94
4000	FDSA	52180.32
5000	SA	67095.93
5000	FDSA	65987.44
5000	FDSA	66494.64
5000	FDSA	66992.74
5000	FDSA	66979
5000	FDSA	66927
6000	SA	83461.9
6000	SA	83461.9
6000	SA	79714.64
6000	SA	79714.64
6000	SA	83461.9
6000	GA	83461.9
6000	GA	83477
6000	GA	83512
6000	GA	83419
6000	GA	83562
6000	FDSA	83429
6000	FDSA	83501
6000	FDSA	83461.9
6000	FDSA	83452.79

DEMAND	ALGORITHM	TRAVEL TIME (MIN)
6000	FDSA	81441.06
13000	SA	312855.9217
13000	SA	387200.7417
13000	SA	402751.3717
13000	SA	251905.2133
13000	SA	277209.9
13000	GA	290596.345
13000	GA	275414.78
13000	GA	276006.4567
13000	GA	279180.9717
13000	GA	276762.755
13000	FDSA	290596.345
13000	FDSA	275414.78
13000	FDSA	276006.4567
13000	FDSA	279180.9717
13000	FDSA	276762.755
14000	SA	294573.5717
14000	SA	324633.895
14000	SA	336440.6733
14000	SA	348480.0017
14000	SA	294573.5717
14000	GA	290450.0467
14000	GA	304179.1867
14000	GA	299597.1667
14000	GA	297015.165
14000	GA	297015.165
14000	FDSA	290450.0467
14000	FDSA	304179.1867
14000	FDSA	299597.1667
14000	FDSA	297015.165
14000	FDSA	297015.165
15000	SA	319565.9133
15000	SA	319565.9133
15000	SA	319565.9133
15000	SA	319565.9133
15000	SA	319565.9133
15000	GA	333229.1583

DEMAND	ALGORITHM	TRAVEL TIME (MIN)
15000	GA	333229.1583
15000	GA	333229.1583
15000	GA	333229.1583
15000	GA	333229.1583
15000	FDSA	333229.1583
15000	FDSA	333229.1583
15000	FDSA	333229.1583
15000	FDSA	333229.1583
15000	FDSA	333229.1583
16000	SA	343296.0533
16000	SA	346309.0417
16000	SA	337040.9617
16000	SA	353023.0383
16000	SA	353041.2717
16000	GA	342491.2967
16000	GA	343681.685
16000	GA	340587.28
16000	GA	346185.8567
16000	GA	333229.1583
16000	FDSA	342491.2967
16000	FDSA	343681.685
16000	FDSA	340587.28
16000	FDSA	346185.8567
16000	FDSA	333229.1583

Appendix B: Akcelik Travel Time Constants

Akcelik Table of Travel Time Constants

		AKCELIK'S TIME-DEPENDENT	
		FOR FIGURE 2	
		TRAVEL TIME	
DEGREE OF SATN		(S/KM)	
$x = q/Q$	$z = x-1$	t	t/to
		Akcelik ()	Akcelik ()
0.01	−0.990	45.0	1.0
0.25	−0.750	45.6	1.0
0.30	−0.700	45.8	1.0
0.35	−0.650	46.0	1.0
0.40	−0.600	46.2	1.0
0.45	−0.550	46.5	1.0
0.50	−0.500	46.8	1.0
0.55	−0.450	47.2	1.0
0.60	−0.400	47.7	1.1
0.65	−0.350	48.3	1.1
0.70	−0.300	49.2	1.1
0.75	−0.250	50.3	1.1
0.80	−0.200	52.1	1.2
0.85	−0.150	54.8	1.2
0.90	−0.100	60.0	1.3
0.95	−0.050	71.4	1.6
0.96	−0.040	75.4	1.7
0.97	−0.030	80.2	1.8
0.98	−0.020	86.2	1.9
0.99	−0.010	93.3	2.1
1.00	**0.000**	**101.9**	**2.265**
1.01	0.010	111.9	2.5

DEGREE OF SATN		AKCELIK'S TIME-DEPENDENT FOR FIGURE 2 TRAVEL TIME (S/KM)	
1.02	0.020	123.2	2.7
1.03	0.030	135.8	3.0
1.04	0.040	149.3	3.3
1.05	0.050	163.7	3.6
1.06	0.060	178.7	4.0
1.07	0.070	194.2	4.3
1.08	0.080	210.2	4.7
1.13	0.130	293.7	6.5
1.18	0.180	380.4	8.5
1.23	0.230	468.4	10.4
1.28	0.280	557.1	12.4
1.33	0.330	646.2	14.4
1.38	0.380	735.5	16.3
1.43	0.430	824.9	18.3

(*Source:* www.sidrasolutions.com)

Appendix C: Matlab Output Plots

MATLAB PLOTS AND OUTPUT

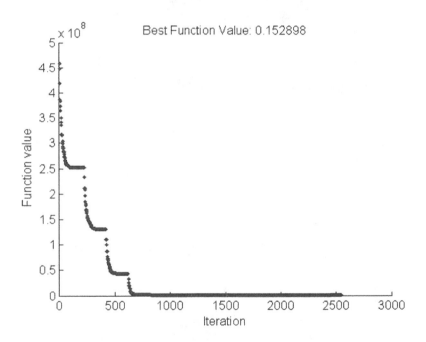

Simulated Annealing Plot D=950 veh/hr

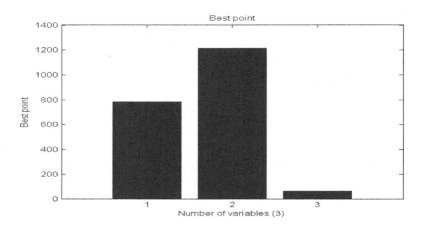

Simulated Annealing Solution D=2000vph

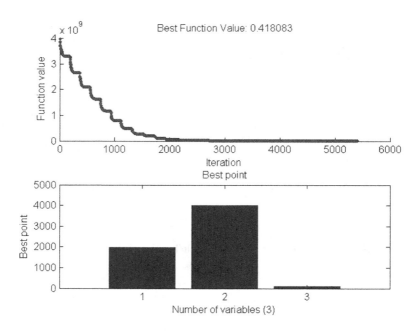

$$\text{Demand} = 6000, \text{Cap}\left(\text{EL}\right) = 4000, \text{density} = 41$$

$$t1 = TT + 0.0625 * ((x-1) + ((x-1)^2 + (0.0016 * (x)))^0.5)$$

$$t1 - TT = toll / VOT = 0.02134 ==> toll = 0.02134 * 20$$
$$= \$0.4269 ====> add.Toll = \$0.50 \, Toll = \$.50 + \$.50 = \$1.00$$

DEMAND	CAPACITY (EL)	OPTIMAL HOT FLOW (X1)	OPTIMAL GP FLOW (X2)	DENSITY	BASE TOLL
6000	4000	1988	4012	41	$1.00

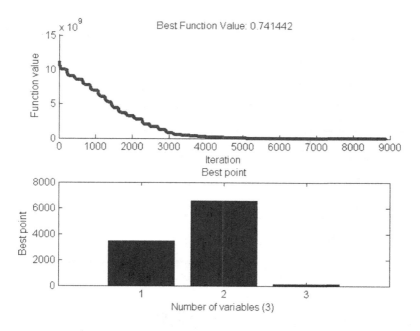

$$Demand = 10000 \; Cap(EL) = 4000, density = 75 t1 = TT + 0.0625 * ((1-1)$$
$$+ ((1-1)^2 + (0.0016 * (x)))^0.5)$$

$$Total \; toll = \$.50 + \$1.25 = \$1.75$$

DEMAND	CAPACITY (EL)	OPTIMAL HOT FLOW (X1)	OPTIMAL GP FLOW (X2)	DENSITY	BASE TOLL
10000	4000	3475	6525	75	$1.75

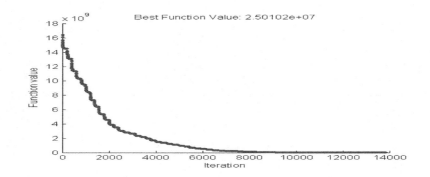

DEMAND	CAPACITY	OPTIMAL HOT FLOW (X1)	OPTIMAL GP FLOW (X2)	DENSITY	BASE TOLL
12000	4000	4000	8000	87	$4.00

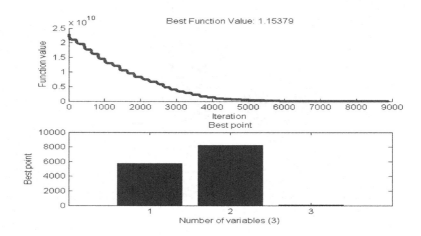

DEMAND	CAPACITY (EL)	OPTIMAL HOT FLOW (X1)	OPTIMAL GP FLOW (X2)	DENSITY	BASE TOLL
14000	4000	5770	8230	125	$7.00 (max)

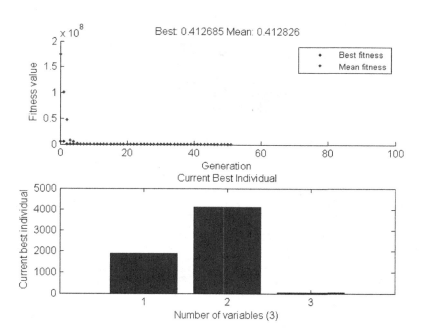

DEMAND	CAPACITY (EL)	OPTIMAL HOT FLOW (X1)	OPTIMAL GP FLOW (X2)	DENSITY	BASE TOLL
6000	4000	1892	4108	41	$1.00

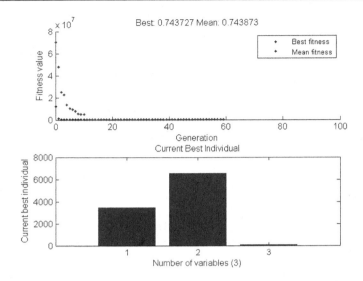

DEMAND	CAPACITY	OPTIMAL HOT FLOW (X1)	OPTIMAL GP FLOW (X2)	DENSITY	BASE TOLL
10000	4000	3460	6540	75	$1.50

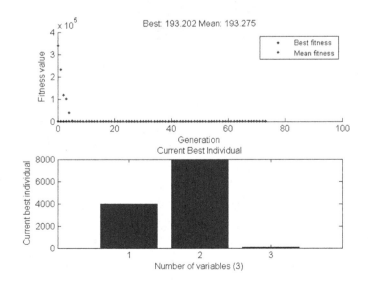

DEMAND	CAPACITY (EL)	OPTIMAL HOT FLOW (X1)	OPTIMAL GP FLOW (X2)	DENSITY	BASE TOLL
12000	4000	4000	8000	87	$3.75

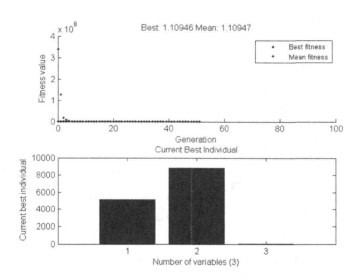

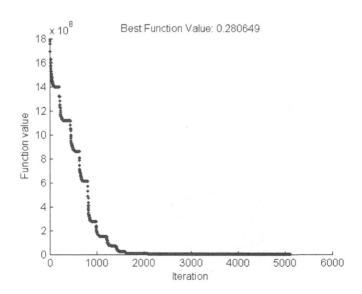

Simulated Annealing Algorithm Plot D=4000vph

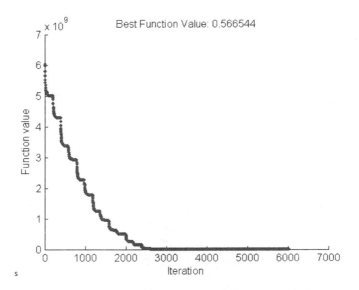

s

Simulated Annealing Algorithm Plot d=7000vph

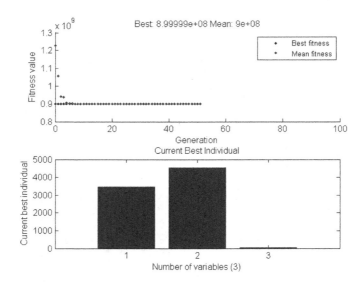

Genetic Algorithm Plot and Solution D=8000vph

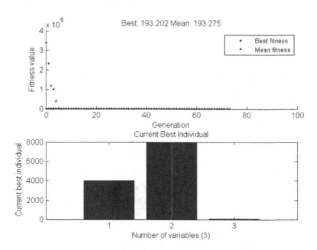

Genetic Algorithm Plot D and Solution =12000vph

DEMAND	CAPACITY (EL)	OPTIMAL HOT FLOW (X1)	OPTIMAL GP FLOW (X2)	DENSITY	BASE TOLL
14000	4000	5175	8825	125	$7.00 (max)

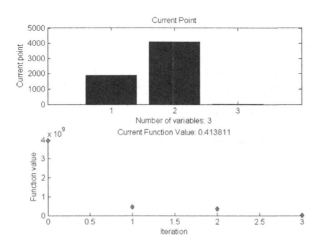

DEMAND	CAPACITY (EL)	OPTIMAL HOT FLOW (X1)	OPTIMAL GP FLOW (X2)	DENSITY	BASE TOLL
6000	4000	1800	4200		$1.00

Genetic Algorithm Plot D and Solution D=6000

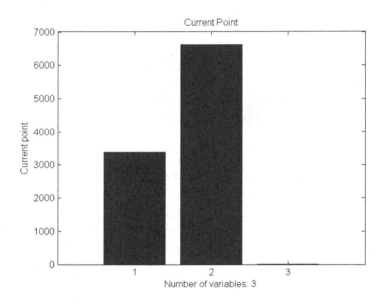

DEMAND	CAPACITY (EL)	OPTIMAL HOT FLOW (X1)	OPTIMAL GP FLOW (X2)	DENSITY	BASE TOLL
10000	4000	3375	6625	75	$1.50

FDSA Plot and Solution D= 10,000

s

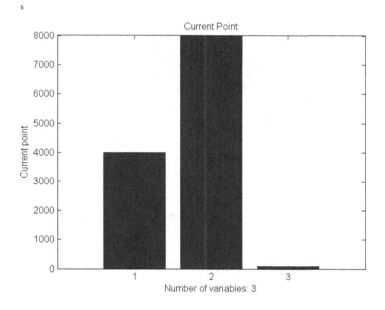

DEMAND	CAPACITY (EL)	OPTIMAL HOT FLOW (X1)	OPTIMAL GP FLOW (X2)	DENSITY	BASE TOLL
12000	4000	4000	8000	87	$3.75

FDSA Plot and Solution D= 12,000

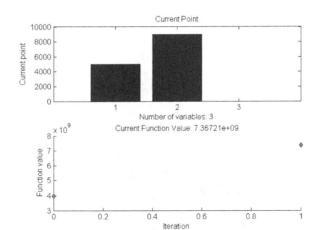

DEMAND	CAPACITY (EL)	OPTIMAL HOT FLOW (X1)	OPTIMAL GP FLOW (X2)	DENSITY	BASE TOLL
14000	4000	5175	8825	125	$7.00 (max)

Genetic Algorithm Plot and Solution, D=14000

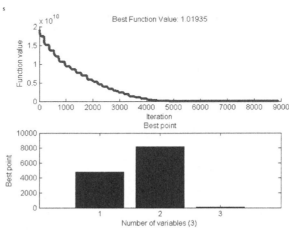

Simulated Annealing Plot, D=13000

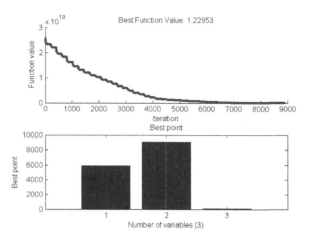

Simulated Annealing Plot, D=15000

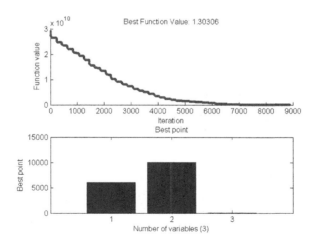

Simulated Annealing Plot, D= 16000

Appendix D: Value of Time Constants

Value of Travel Time Table (source: USDOT)

CATEGORY	SURFACE MODES* *(EXCEPT HIGH-SPEED RAIL)*	AIR AND HIGH-SPEED RAIL TRAVEL
Local Travel—Personal BusinessIntercityTravel-Personal Business	$23.90 $22.90 $23.90 $22.90	$45.60 $57.20

Truck Drivers	$24.70
Bus Drivers	$24.50
Transit Rail Operators	$40.40
Locomotive engineers	$34.30
Airline Pilots and Engineers	$76.10

Appendix E: Station Data I-95 Hot Lane

Hourly Demand on I-95 Express Oct 27, 2012

I-95 HOT On-ramp to I-95 NB at NW 46 St (MM 5.20)	

Station:

#	DATE	TIME	
1	10/27/2012	0:00:00	148
2	10/27/2012	0:15:00	131
3	10/27/2012	0:30:00	105
4	10/27/2012	0:45:00	92
5	10/27/2012	1:00:00	84
6	10/27/2012	1:15:00	74
7	10/27/2012	1:30:00	56
8	10/27/2012	1:45:00	72
9	10/27/2012	2:00:00	67
10	10/27/2012	2:15:00	56
11	10/27/2012	2:30:00	56
12	10/27/2012	2:45:00	53
13	10/27/2012	3:00:00	54
14	10/27/2012	3:15:00	36
15	10/27/2012	3:30:00	40
16	10/27/2012	3:45:00	55
17	10/27/2012	4:00:00	46
18	10/27/2012	4:15:00	46

#	DATE	TIME	
19	10/27/2012	4:30:00	36
20	10/27/2012	4:45:00	41
21	10/27/2012	5:00:00	46
22	10/27/2012	5:15:00	55
23	10/27/2012	5:30:00	51
24	10/27/2012	5:45:00	56
25	10/27/2012	6:00:00	63
26	10/27/2012	6:15:00	75
27	10/27/2012	6:30:00	85
28	10/27/2012	6:45:00	87
29	10/27/2012	7:00:00	99
30	10/27/2012	7:15:00	123
31	10/27/2012	7:30:00	136
32	10/27/2012	7:45:00	161
33	10/27/2012	8:00:00	144
34	10/27/2012	8:15:00	179
35	10/27/2012	8:30:00	185
36	10/27/2012	8:45:00	209
37	10/27/2012	9:00:00	216
38	10/27/2012	9:15:00	270
39	10/27/2012	9:30:00	271
40	10/27/2012	9:45:00	259
41	10/27/2012	10:00:00	289
42	10/27/2012	10:15:00	311
43	10/27/2012	10:30:00	323
44	10/27/2012	10:45:00	340
45	10/27/2012	11:00:00	323
46	10/27/2012	11:15:00	355
47	10/27/2012	11:30:00	352
48	10/27/2012	11:45:00	370
49	10/27/2012	12:00:00	344
50	10/27/2012	12:15:00	355

#	DATE	TIME	
51	10/27/2012	12:30:00	377
52	10/27/2012	12:45:00	340
53	10/27/2012	13:00:00	356
54	10/27/2012	13:15:00	376
55	10/27/2012	13:30:00	349
56	10/27/2012	13:45:00	331
57	10/27/2012	14:00:00	320
58	10/27/2012	14:15:00	318
59	10/27/2012	14:30:00	364
60	10/27/2012	14:45:00	337
61	10/27/2012	15:00:00	333
62	10/27/2012	15:15:00	367
63	10/27/2012	15:30:00	347
64	10/27/2012	15:45:00	305
65	10/27/2012	16:00:00	354
66	10/27/2012	16:15:00	327
67	10/27/2012	16:30:00	295
68	10/27/2012	16:45:00	305
69	10/27/2012	17:00:00	331
70	10/27/2012	17:15:00	323
71	10/27/2012	17:30:00	278
72	10/27/2012	17:45:00	342
73	10/27/2012	18:00:00	318
74	10/27/2012	18:15:00	352
75	10/27/2012	18:30:00	330
76	10/27/2012	18:45:00	278
77	10/27/2012	19:00:00	270
78	10/27/2012	19:15:00	258
79	10/27/2012	19:30:00	271
80	10/27/2012	19:45:00	252
81	10/27/2012	20:00:00	249
82	10/27/2012	20:15:00	234

#	DATE	TIME	
83	10/27/2012	20:30:00	203
84	10/27/2012	20:45:00	192
85	10/27/2012	21:00:00	205
86	10/27/2012	21:15:00	179
87	10/27/2012	21:30:00	203
88	10/27/2012	21:45:00	208
89	10/27/2012	22:00:00	224
90	10/27/2012	22:15:00	212
91	10/27/2012	22:30:00	205
92	10/27/2012	22:45:00	216
93	10/27/2012	23:00:00	179
94	10/27/2012	23:15:00	209
95	10/27/2012	23:30:00	152
96	10/27/2012	23:45:00	165

References

Al-Deek, H., A. Mohamed, and A. E. Radwan. "A New Model for the Evaluation of Traffic Operations at Electronic Toll Collection Plazas," Journal of the Transportation Research Board, (1710), December 2000.

Al-Deek, H., A. E. Radwan, A. Mohammed, and J. Klodzinski. "Evaluating the Improvements of Traffic Operations at a Real-Life Toll Plaza with Electronic Toll Collection," ITS Journal, vol. 3(3), 1996.

Aleskerov, F., D. Bouyssou, and B. Monjardet. "Utility Maximization Choice Preference. Studies in Economic Theory." September 2007.

Amemiya, T. "The Maximum Likelihood, the Minimum Chi Square, and the Non-Linear Weighted Least Squares Estimator in the General Qualitative Response Model," Journal of the American Statistical Association, vol. 71, 1976, pp. 347–351.

Atherton, T., and M. Ben-Akiva. "Transferability and Updating of Disaggregated Travel Demand Models," Paper Presented at the 55th Annual Meeting of the Transportation Research Board, 1976, Washington, DC.

Bagley, M. N., J. S. Mannering, and P. L. Mokhtarian. Telecommuting Centers and Related Concepts: A Review of Practice (UCD-ITS-RR-94-4). Institute of Transportation Studies, University of California, Davis, March 1994.

Ben-Akiva, M., and S. Lerman. Discrete Choice Analysis Theory and Application to Travel Demand (edited by Marvin Manheim). MIT Press Series in Transportation Studies, December 1985.

Ben-Akiva, M., and M. Richards. A Disaggregate Travel Demand Model. Saxon House, University of Michigan, 1975.

Benson, H., D. Shanno, and R. Vanderbei. Interior-Point Methods for Non-Convex Nonlinear Programming: Jamming and Comparative Numerical Testing (ORFE-00-02). Operations Research and Financial Engineering, Princeton University, August 28, 2000.

Bernardino, A., M. Ben-Akiva, and I. Salomon. "A Stated Preference Approach to Modeling the Adoption of Telecommuting," Transportation Research Record, vol. 1413, 1993, pp. 22–30.

Bertsikas, D. Non-Linear Programming. 2nd ed. Athena Scientific, 1999.

Börjesson, M., and I. Kristoffersson. "The Swedish Congestion Charges: Ten Years on: and Effects of Increasing Charging Levels," Working Papers in Transport Economics 2017:2, 2017. CTS—Centre for Transport Studies Stockholm (KTH and VTI).

Brands, T., C. Goudappel, V. E. Berkum, and D. Hendrik. "Optimal Toll Design in Dynamic Traffic Networks Using a Pattern Search Approximation Algorithm," Transportation Research Board 88th Annual Meeting, January 11–15, 2009, Washington, DC. Transportation Research Board.

Burris, M. W. "The Toll-Price Component of Travel Demand Elasticity," International Journal of Transport Economics / Rivista Internazionale Di Economia Dei Trasporti, vol. 30(1), 2003, pp. 45–59.

Button, K., and E. Verhoef. Road Pricing Traffic Congestion and the Environment: Issues of Efficiency and Social Feasibility, 1998.

Carson, J. Monitoring and Evaluating Managed Lane Facility Performance. Texas Transportation Institute, September 2005. http://tti.tamu.edu/documents/0-4160-23.pdf.

Chandra, R. S., and H. Al-Deek. "Prediction of Freeway Traffic Speeds and Volumes Using Vector Autoregressive Models," Journal of Intelligent Transportation Systems (J-ITS), vol. 13(2), April–June 2009, pp. 53–72.

Daganzo, C. F. "The Cell-Transmission Model. Part I: A Simple Dynamic Representation of Highway Traffic," California PATH Research Report, UCB-ITS-PRR-93-7, Institute of Transportation Studies, University of California, Berkeley, 1993.

Daly, A. and S. Zachary. Improved multiple choice models, in: D.Hensher and Q. Dalvi, eds., Identifying and measuring the determinants of mode choice (Teakfield, London), 1979, pp. 335–357.

Daniel, J., and M. Pahwa. "Comparison of Three Empirical Models of Airport Congestion Pricing," Journal of Urban Economics, vol. 47(1), January 2000, pp. 1–38.

DeCorla-Souza, P. T. "A Concept of Operations for Peak-Period Pricing on Metropolitan Freeway Systems," Transportation Research Board 88th Annual Meeting, January 11–15, 2009, Washington DC. Transportation Research Board.

De Palma, A., M. Kilani, and R. Lindsey. "Congestion Pricing on a Road Network: A Study Using the Dynamic Equilibrium Simulator METROPOLIS," Transportation Research, Part A: Policy and Practice, vol. 39(7–9), 2005, pp. 588–611.

De Palma, A., and R. Lindsey. "Private Roads, Competition, and Incentives to Adopt Time-Based Congestion Tolling," Journal of Urban Economics, vol. 52(2), 2002, pp. 217–241.

De Palma, A., and R. Lindsey. "Congestion Pricing with Heterogeneous Travelers: A General-Equilibrium Welfare Analysis," Networks and Spatial Economics, vol. 4(2), 2004, pp. 135–160.

DeSanctis, G. "Attitudes Toward Telecommuting: Implications for Work-at-Home Programs," Information and Management, vol. 7, 1984, pp. 133–139.

Eliasson, J., L. Hultkrantz, L. Nerhagen, L. S. Rosqvist. "The Stockholm Congestion—Charging Trial 2006: Overview of Effects," Transportation Research Part A: Policy and Practice, vol. 43(3), 2009, pp. 240–250.

Evaluation Plan Framework for 95 Express Managed Lanes. Cambridge Systematics, Inc, March 30, 2009.

Federal Highway Administration (FHWA). "Examining Congestion Pricing. Implementation Issues. Searching for Solutions: A Policy Discussion," Series No. 6, 1992.

Fishburn, P. Nonlinear Preference and Utility Theory. Johns Hopkins University Press, 1988.

Fishburn, P. "The Foundations of Expected Utility (Theory and Decision Library)," December 23, 2010.

Franklin, J. "The Role of Context in the Equity Effects of Congestion Pricing," Transportation Research Board 91st Annual Meeting, January 22–26, 2012, Washington DC. Transportation Research Board.

Friesz, T. L. Dynamic Optimization and Differential Games. Springer, 2007.

Gardner, L., A. Unnikrishnan, and S. Waller. "Robust Pricing of Transportation Networks Under Uncertain Demand," Transportation Research Record: Journal of the Transportation Research Board, (2085), 2010.

Graham, D. P., S. Datta, N. Sagar, P. Doody, R. Slone, and O.-P. Hilmola. "Forecasting and Risk Analysis in Supply Chain Management: GARCH Proof of Concept," in Wu and Blackhurst, eds., Managing Supply Chain Risk and Vulnerability. Springer, 2009, pp. 187–207.

Hartman, R. I., C. R. Stoner, and R. Arora. "An Investigation of Selected Variables Affecting Telecommuting Productivity and Satisfaction," Journal of Business and Psychology, vol. 6(2), 1991, pp. 207–225.

Hultgren, L., and K. Kawada. "San Diego's Interstate 15 High-Occupancy/Toll Lane Facility Using Value Pricing," Institute of Transportation Engineers Journal, vol. 69(6), June 1999, p. 27.

Iseki, H., T. Brian, and D. Alexander. "Examining the Linkages between Electronic Roadway Tolling Technologies and Road Pricing Policy Objectives," Transportation Research Board 89th Annual Meeting, January 10–14, 2010, Washington, DC. Transportation Research Board.

Jonas Eliasson, Lars Hultkrantz, Lena Nerhagen, Lena Smidfelt Rosqvist. The Stockholm congestion—charging trial 2006: Overview of effects, Transportation Research Part A: Policy and Practice, 43(3), 2009, pp. 240–250, ISSN 0965-8564.

Jang, K., and K. Chung. "A Dynamic Congestion Pricing Strategy for High-Occupancy Toll Lanes," Transportation Research Board 89th Annual Meeting, January 1–14, 2010, Washington, DC. Transportation Research Board.

Kleinman, N. L., S. D. Hill, and V. A. Ilenda. "SPSA/SIMMOD Optimization of Air Traffic Delay Cost," Proceedings of the 1997 American Control Conference (Cat. No. 97CH36041), Albuquerque, NM, vol. 2, 1997, pp. 1121–1125. doi:10.1109/ACC.1997.609707.

Kleinman, N. L., J. C. Spall, and D. Q. Naiman. "Simulation-Based Optimization with Stochastic Approximation Using Common Random Numbers," Management Science, vol. 45 (11), 1999, pp. 1570–1578.

Klodzinski, J., and Adler, T. "Application for Post Processing of Travel Demand Forecasts for Estimating Express Lane Traffic and Variable Toll Rates," Transportation Research Board 89th Annual Meeting, January 1–14, 2010, Washington, DC. Transportation Research Board.

Labi, S., and A. Issariyanukula. "Financial and Technical Feasibility of Dynamic Congestion Pricing as a Revenue Generation Source in Indiana – Exploiting the Availability of Real-Time Information and Dynamic Pricing Technologies," USDOT Region V Regional University Transportation Center Final Report (Nextrans Project No. 044PY02), October 19, 2011. Purdue University.

Leape, J. "The London Congestion Charge," Journal of Economic Perspectives, vol. 20(4), 2006, pp. 157–176.

Lee, K.-S., and A. Hobeika. "Application of Dynamic Value Pricing Through Enhancements to TRANSIMS," Transportation Research Record, vol. 2003, 2007, pp. 7–16. doi:10.3141/2003-02.

Lewis, N. C. Road Pricing: Theory and Practice. Thomas Telford, 1993.

Lo, H., M. Hickman, and M. Walstad. "An Evaluation Taxonomy for Congestion Pricing," Research Reports, California Partners for Advanced Transit and Highways (PATH) (Report UCB-ITS-PRR-96-10), Institute of Transportation Studies (UCB), UC Berkeley, January 1, 1996.

Lu, C., and H. Mahmassani. "Dynamic Pricing with Heterogeneous Users: Gap-Driven Solution Approach for Bicriterion Dynamic User Equilibrium Problem," Transportation Research Record: Journal of the Transportation Research Board, vol. 2090, 2010.

Lu, C., and H. S. Mahmassani. "Modeling User Responses to Pricing: Simultaneous Route and Departure Time Network Equilibrium with Heterogeneous Users," Journal of the Transportation Research Board, vol. 2085, 2008.

Mahmassani, H., X. Zhou, and C.-C. Lu. "Toll Pricing and Heterogeneous Users: Approximation Algorithms for Finding Bicriterion Time-Dependent Efficient Paths in Large-Scale Traffic Networks," Transportation Research Record: Journal of the Transportation Research Board, vol. 1923, 2005.

Manski, C. "Maximum Score Estimation of the Stochastic Utility Model of Choice," Journal of Econometrics, vol. 3, 1975, pp. 205–228.

Manski, C., and S. Lerman. "The Estimation of Choice Probabilities from Choice-Based Samples," Econometrica, 1976.

Manski, C., and D. McFadden. "Alternative Estimators and Sample Econometric Models for Probabilistic Choice among Products," Journal of Business, vol. 53(3), 1980.

Melo, E. "Congestion Pricing and Learning in Traffic Network Games," Journal of Public Economic Theory, vol. 13(3), 2011, pp. 351–367.

Michalaka, D., Y. Lou, and Y. Yin. "Proactive and Robust Dynamic Pricing Strategies for High-Occupancy-Toll (HOT) Lanes," Transportation Research Board 90th Annual Meeting, January 23–27, 2011, Washington, DC. Transportation Research Board.

Modi, V., Y. Yin, S. Washburn, D. Wu, A. Kulshrestha, D. Michalaka, and J. Lu. "Managed Lane Operations-Adjusted Time of Day Pricing vs Near Real Time Dynamic Pricing," FDOT Contract BDK77_977-04, February 2012.

Mokhtarian, P., and I. Salomon. "Modeling the Choice of Telecommuting: Setting the Context," Environment and Planning A, vol. 26(5), 1994, pp. 749–766.

Morgul, E. F., and K. Ozbay. "Simulation-Based Evaluation of a Feedback Based Dynamic Congestion Pricing Strategy on Alternate Facilities," Transportation Research Board 90th Annual Meeting, January 23–27, 2011, Washington, DC. Transportation Research Board.

Ohazulike, A., C. J. Michiel, G. Still, and E. C. Van Berkum. "Multi-Objective Road Pricing: A Game Theoretic and Multi-Stakeholder Approach," Transportation Research Board 91st Annual Meeting, January 22–26, 2012, Washington, DC. Transportation Research Board.

Ozbay, K., E. F. Morgul, S. Ukkusuri, V. Iyer, S. Yushimito, and W. Fernando. "A Mesoscopic Simulation Evaluation of Dynamic Congestion Pricing Strategies for New York City Crossings," Transportation Research Board 90th Annual Meeting, January 23–27, 2011, Washington, DC. Transportation Research Board.

Pels, E., and E. T. Verhoef. "The Economics of Airport Congestion Pricing," Journal of Urban Economics, vol. 55, 2004, pp. 257–277.

Perez, B. Evaluation and Performance Measurement of Congestion Pricing Projects. Transportation Research Board, 2011.

Radwan, A. E., and J. Selter. "The Science of Modeling and Simulation of Vehicular Traffic," The Society for Modeling and Simulation International, Summer Computer Conference, 2002, San Diego, CA.

Small, K., and J. Gomez-Ibañez. Road Pricing for Congestion Management: The Transition from Theory to Policy. The University of California Transportation Center, University of California at Berkeley, 1998, pp. 213–227.

Small, K., and E. Verhoef. The Economics of Urban Transportation. Routledge, 2007.

Spall, J. C. "A Stochastic Approximation Technique for Generating Maximum Likelihood Parameter Estimates," Proceedings of the American Control Conference, Minneapolis, MN, June 1987, pp. 1161–1167.

Spears, S., M. G. Boarnet, and S. Handy. Draft Policy Brief on the Impacts of Road User Pricing Based on a Review of the Empirical Literature, for Research on Impacts of Transportation and Land Use-Related Policies. California Air Resources Board.

Sullivan, E. "Continuation Study to Evaluate the Impacts of the SR 91 Value-Priced Express Lanes," Final Report, Cal Poly State University, Department of Civil and Environmental Engineering, Applied Research and Development Facility, December 2000. www.vta.org/expresslanes/pdf/cal_poly_exp_lanes_sr91_2.pdf

Sullivan, E., and M. Burris. "Benefit-Cost Analysis of Variable Pricing Projects: SR-91 Express Lanes," Journal of Transportation Engineering, vol. 132(3), 2006, pp. 191–197.

Tuerk, A., M. MacDonald, D. Graham, and H. Li. "Impact of Congestion Charging on Traffic Volumes: Evidence from London," Transportation Research Board 91st Annual Meeting, January 22–26, 2012, Washington, DC. Transportation Research Board.

Ukkusuri, S., K. S. Ampol, T. Waller, and K. Kockelman. "Congestion Pricing Technologies: A Comparative Evaluation," in F. N. Gustavsson, ed. New Transportation Research Progress. Nova Science Publishers, 2008, pp. 121–142 (Presented at the 84th Annual Meeting of the Transportation Research Board, January 2005).

Vaze, V., and C. Barnhart. "Airline Frequency Competition in Airport Congestion Pricing," Transportation Research Record: Journal of the Transportation Research Board, vol. 2266, 2012.

Verhoef, E. Pricing in Road Transport: A Multidisciplinary Perspective, 2008.

Verhoef, E. T. Second-Best Congestion Pricing in General Networks-Algorithms for Finding Second-Best Optimal Toll Levels and Toll Points (Tinbergen Institute Discussion Papers 00-084/3). Tinbergen Institute, 2000.

Wadoo, S., P. Kachroo, K. Ozbay, and S. Neveen. "Feedback Based Dynamic Congestion Pricing," Transportation Research Board 90th Annual Meeting, January 23–27, 2011, Washington, DC. Transportation Research Board.

Xu, S. "Development and Test of Dynamic Congestion Pricing Model," SM Thesis, Massachusetts Institute of Technology, 2009.

Yang, H. "Optimal Road Tolls Under Conditions of Queuing and Congestion," Transportation Research Part A: Policy and Practice, vol. 30(5), September 1996, pp. 319–332.

Yang, H., X. Wei, H. Bing-Sheng, and M. Qiang. "Road Pricing for Congestion Control with Unknown Demand and Cost Functions," Transportation Research Part C: Emerging Technologies, vol. 18(2), April 2010, pp. 157–175.

Yang, L., and C. Chu. "Stochastic Model for Traffic Flow Prediction and Its Validation," Transportation Research Board 90th Annual Meeting, January 23–27, 2011, Washington, DC. Transportation Research Board.

Yang, L., R. Saigal, and H. Zhou. "Distance-Based Dynamic Pricing Strategy for Managed Toll Lanes," Transportation Research Board 91st Annual Meeting, January 22–26, 2012, Washington, DC. Transportation Research Board.

Zarrillo, M., and A. E. Radwan. "Modeling the OOCEA's Toll Network of Highways Using Plaza Capacity Analyses," The 8th World Congress on Intelligent Transportation Systems, 2001, Sydney.

Zarrillo, M., and A. E. Radwan. "Identification of Bottlenecks on a Toll Network of Highways," 7th International ASCE Specialty Conference on Applications of Advance Technology in Transportation, August 2002, Boston, MA.

Zarrillo, M. L., A. E. Radwan, and H. M. Al-Deek. "Modeling Traffic Operations at Electronic Toll Collection and Traffic Management Systems," The International Journal on Computers and Industrial Engineering, vol. 33(3–4), December 1997, pp. 857–886.

Zarrillo, M. L., A. E. Radwan, and J. Dowd. "Toll Network Capacity Calculator: Operations Management and Assessment Tool for Toll Network Operators," Transportation Research Board Journal, vol. 1781, May 2002.

Zheng, N., R. Axhausen, W. Kay, and N. Geroliminis. "A Dynamic Cordon Pricing Scheme Combining Macroscopic and Agent-Based Traffic Models," Transportation Research Board 91st Annual Meeting, January 22–26, 2012, Washington, DC. Transportation Research Board.

Zhong, S. "Dynamic Congestion Pricing for Multi-Class and Multi-Modes Transportation Systems with Asymmetric Cost Functions," Transportation Research Board 88th Annual Meeting, January 11–15, 2009, Washington, DC. Transportation Research Board.

Zhong, S., M. Bushell, and W. Deng. "A Reliability-Based Stochastic System Optimum Congestion Pricing Model Under ATIS with Endogenous Market Penetration and Compliance Rate," Transportation Research Board 91st Annual Meeting, January 22–26, 2012, Washington, DC. Transportation Research Board.

Index

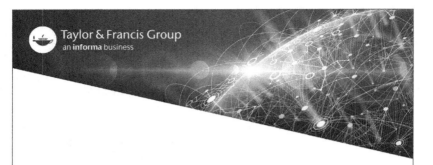